Science, Technology and Medicine in Modern History

General Editor: **John V. Pickstone**, Centre for the History of Science, Technology and Medicine, University of Manchester, England (www.man.ac.uk/CHSTM).

One purpose of historical writing is to illuminate the present. At the start of the third millennium, science, technology and medicine are enormously important, yet their development is little studied.

The reasons for this failure are as obvious as they are regrettable. Education in many countries, not least in Britain, draws deep divisions between the sciences and the humanities. Men and women who have been trained in science have too often been trained away from history, or from any sustained reflection on how societies work. Those educated in historical or social studies have usually learned so little of science that they remain thereafter suspicious, overawed, or both.

Such a diagnosis is by no means novel, nor is it particularly original to suggest that good historical studies of science may be peculiarly important for understanding our present. Indeed this series could be seen as extending research undertaken over the last half-century. But much of that work has treated science, technology and medicine separately; this series aims to draw them together, partly because the three activities have become ever more intertwined. This breadth of focus and the stress on the relationships of knowledge and practice are particularly appropriate in a series which will concentrate on modern history and on industrial societies. Furthermore, while much of the existing historical scholarship is on American topics, this series aims to be international, encouraging studies on European material. The intention is to present science, technology and medicine as aspects of modern culture, analysing their economic, social and political aspects, but not neglecting the expert content which tends to distance them from other aspects of history. The books will investigate the uses and consequences of technical knowledge, and how it was shaped within particular economic, social and political structures.

Such analyses should contribute to discussions of present dilemmas and to assessments of policy. 'Science' no longer appears to us as a triumphant agent of Enlightenment, breaking the shackles of tradition, enabling command over nature. But neither is it to be seen as merely oppressive and dangerous. Judgement requires information and careful analysis, just as intelligent policy-making requires a community of discourse between men and women trained in technical specialities and those who are not.

This series is intended to supply analysis and to stimulate debate. Opinions will vary between authors; we claim only that the books are based on searching historical study of topics which are important, not least because they cut across conventional academic boundaries. They should appeal not just to historians, nor just to scientists, engineers and doctors, but to all who share the view that science, technology and medicine are far too important to be left out of history.

Titles include:

Julie Anderson, Francis Neary and John V. Pickstone
SURGEONS, MANUFACTURERS AND PATIENTS
A Transatlantic History of Total Hip Replacement

Roberta E. Bivins
ACUPUNCTURE, EXPERTISE AND CROSS-CULTURAL MEDICINE

Roger Cooter
SURGERY AND SOCIETY IN PEACE AND WAR
Orthopaedics and the Organization of Modern Medicine, 1880–1948

Jean-Paul Gaudillière and Ilana Löwy (editors)
THE INVISIBLE INDUSTRIALIST
Manufacture and the Construction of Scientific Knowledge

Ayesha Nathoo
HEARTS EXPOSED
Transplants and the Media in 1960s Britain

Neil Pemberton and Michael Worboys
MAD DOGS AND ENGLISHMEN
Rabies in Britain, 1830–2000

Cay-Rüdiger Prüll, Andreas-Holger Maehle and Robert Francis Halliwell
A SHORT HISTORY OF THE DRUG RECEPTOR CONCEPT

Thomas Schlich
SURGERY, SCIENCE AND INDUSTRY
A Revolution in Fracture Care, 1950s–1990s

Eve Seguin (editor)
INFECTIOUS PROCESSES
Knowledge, Discourse and the Politics of Prions

Crosbie Smith and Jon Agar (editors)
MAKING SPACE FOR SCIENCE
Territorial Themes in the Shaping of Knowledge

Stephanie J. Snow
OPERATIONS WITHOUT PAIN
The Practice and Science of Anaesthesia in Victorian Britain

Carsten Timmermann and Julie Anderson (editors)
DEVICES AND DESIGNS
Medical Technologies in Historical Perspective

Science, Technology and Medicine in Modern History
Series Standing Order ISBN 978–0–333–71492–8 hardcover
Series Standing Order ISBN 978–0–333–80340–0 paperback
(outside North America only)

You can receive future titles in this series as they are published by placing a standing order. Please contact your bookseller or, in case of difficulty, write to us at the address below with your name and address, the title of the series and one of the ISBNs quoted above.

Customer Services Department, Macmillan Distribution Ltd, Houndmills, Basingstoke, Hampshire RG21 6XS, England

---

A Short History of the Drug Receptor Concept

Cay-Rüdiger Prüll
Albert-Ludwigs-Universität Freiburg, Germany

Andreas-Holger Maehle
Durham University, UK

Robert Francis Halliwell
University of the Pacific, USA

First published 2009 by
PALGRAVE MACMILLAN

Palgrave Macmillan in the UK is an imprint of Macmillan Publishers Limited,
registered in England, company number 785998, of Houndmills, Basingstoke,
Hampshire RG21 6XS.

Palgrave Macmillan in the US is a division of St Martin's Press LLC,
175 Fifth Avenue, New York, NY 10010.

Palgrave Macmillan is the global academic imprint of the above companies
and has companies and representatives throughout the world.

Palgrave® and Macmillan® are registered trademarks in the United States,
the United Kingdom, Europe and other countries

ISBN-13: 978–0–230–55415–3   hardback

This book is printed on paper suitable for recycling and made from fully
managed and sustained forest sources. Logging, pulping and manufacturing
processes are expected to conform to the environmental regulations of the
country of origin.

A catalogue record for this book is available from the British Library.

Library of Congress Cataloging-in-Publication Data
Prüll, Cay-Rüdiger.
    A short history of the drug receptor concept / Cay-Rüdiger Prüll,
    Andreas-Holger Maehle, Robert Francis Halliwell.
        p. cm. – (Science, technology, and medicine in modern
    history)
    Includes bibliographical references and index.
    ISBN-13: 978–0–230–55415–3 (hardback : alk. paper)

        1. Drug receptors–History. I. Maehle, Andreas-Holger, 1957– II.
    Halliwell, Robert Francis, 1957– III. Title. IV. Series.
    [DNLM: 1. Receptors, Drug–history. 2. Biomedical Research–
    history. 3. History, 19th Century. 4. History, 20th Century.
    5. Pharmacology–history. QV 11.1 P971s 2008]
    RM301.41.P78 2008
    615'.7–dc22
2008030447

Transferred to digital printing in 2009.

Contents

Preface

In historical studies such as the present one, the research becomes a 'social event' – not simply a critique of archival sources at an office or lab desk. Rather, the character of the questions posed and the subsequent interpretation and discussion of results with colleagues make scientific research a social endeavour. Our work on the history of the receptor concept was no exception.

In addition to many enthusiastic discussions in our own research group, we had many wonderful scholarly conversations and exchanges with researchers at congresses, workshops and seminars. All of these people, friends and colleagues, contributed in some respect to our work through their helpful suggestions and comments on our project. As is often the case, many of these influences are not easily retraceable, but they do leave their mark somewhere in this work and we first wish to acknowledge these unnamed colleagues.

In the paragraphs below we also identify and acknowledge those who helped us in various ways to conduct this project. At the outset we received a Wellcome Trust Project Grant (History of Medicine Project Grant 061819) to enable our study and we are much indebted to the Trust for sponsoring this research. Additional financial support came through a Wellcome Trust Enhancement Award, and through grants from the Rockefeller Archive Center and the GlaxoSmithKline Foundation to carry out research on Paul Ehrlich and Raymond P. Ahlquist.

We would also like to express our gratitude to the librarians and archivists of Durham, Newcastle and Cambridge University Libraries, the Archive of the Stiftung Preussischer Kulturbesitz in Berlin Dahlem, the Archive of the Humboldt-University in Berlin and above all the Contemporary Medical Archive Centre at the Wellcome Library in London for their help and support. Furthermore, we are grateful to Marilee S. Creelan, Head of the Collection Services of the Robert B. Greenblatt Library of the Medical School of Georgia at Augusta, Georgia, and to David Stapleton, Director of the Rockefeller Archive Center at Tarrytown, New York, who was especially helpful in our research on Paul Ehrlich, as were Professor Fritz Soergel, Head of the Institute for Biomedical and Pharmaceutical Research in Nuremberg-Heroldsberg and Hans Schadewaldt, Professor Emeritus of the Institute for the History of Medicine at Düsseldorf University.

Fritz Lembeck, Professor Emeritus of Pharmacology at the University of Graz, and Klaus Starke, Professor Emeritus of Pharmacology at the University of Freiburg, both gave important information on Otto Loewi and Walther Straub. David Hazel Clark, Cambridge, provided key information on his father, Alfred Joseph Clark. The chapter on the work of Raymond P. Ahlquist relies very much on interviews with Richard E. White PhD, former student of Ahlquist, now Associate Professor of the Department of Pharmacology and Toxicology at the Medical College of Georgia, Augusta; Lois T. Ellison, MD, Medical Historian in Residence, Provost Emeritus, Professor Emeritus of Medicine and Surgery, formerly Director of the Cardiopulmonary Lab of the Medical College of Georgia, Augusta; the late Armand M. Karow, PhD, Assistant Professor since 1968 and Associate Professor at the Pharmacology Department since 1971; and Jerry J. Buccafusco, since 1979 assistant at the Department, currently Director of the Alzheimer's Research Centre, Professor of Pharmacology and Toxicology, Professor of Psychiatry and Health Behavior, Medical College of Georgia, Augusta. We are grateful for the materials as well as the information they all provided. Concerning these interviews, we are especially indebted to Lowell Greenbaum, the successor of Ahlquist in the Chair of Pharmacology and Toxicology at the Medical School of Georgia, and his wife Gloria, who established the contacts with the interviewees and also supplied us with materials and personal reminiscences on Ahlquist. Thanks to all of them and also to the Nobel Laureate Professor Sir James Black for additional information on Ahlquist's work.

Also, we express our thanks to the research group of Judy Slinn at Oxford Brookes University for valuable comments on the trends in pharmacology over the twentieth century. Tilly Tansey, Wellcome Trust Centre for the History of Medicine at University College London, supported our research in London and contributed, with her expertise, to our work on the Nobel Laureate Sir Henry Hallett Dale.

The last decisive steps in preparing this book were supported by Viviane Quirke, Oxford Brookes University, and John Parascandola, former President of the American Association for the History of Medicine and an internationally recognized expert on the history of pharmacology, who read the manuscript and delivered important advice and help. Thanks are also due to Danny Knapp and Katherine Smith for checking the language and also Elisabeth Ahner for her work on style and layout. Last but not least, we would like to express our gratitude to John Pickstone, editor of the series, for his support and advice, and Michael Strang, Ruth Ireland and Barbara Slater from Palgrave

Macmillan, for their kind cooperation in publishing the outcome of all these efforts.

C.-R. Prüll (Freiburg),
A.-H. Maehle (Durham),
R. F. Halliwell (California)
October 2008

Introduction

> Generally speaking, until really quite recently – well into the
> 20th century in fact – treatment by most available medicines
> was at best only marginally beneficial and at worst positively
> harmful.
>
> (William C. Bowman, 1999)

This book deals with the concept of receptors – a fundamental idea in
science and medicine. Receptors are defined as proteins at the cell surface and within cells that mediate the effect of chemical messengers and
hormones and the actions of many drugs in the body.[1] Although this
concept is approximately 100 years old, it was not until the 1960s that
it became fully accepted and exploited in the scientific community.

The receptor concept is one of those ideas in biomedical sciences which
has had a great impact. Humans have utilized plant and other natural
extracts as medicines to alleviate pain and illness for millennia. Yet, Sir
Henry Dale, as a young medical student at the turn of the twentieth
century, could describe his great disappointment when he realized just
how few (perhaps fewer than 20) useful drugs were available to him,
and how little was known about how even the effective ones worked.[2]
Many of these drugs, such as amyl nitrite, atropine, digitalis, ephedrine,
cocaine, morphine, physostigmine, quinine and salicylates, were in fact
of ancient origin.

Since its introduction, the concept of receptors has served as a scientific basis for understanding how such drugs act in the body and has
provided a significant impetus to the development of new drugs to target these receptors. This fits the argument, as proposed by Drews, that
drug research has contributed more to the progress of medicine during
the past century than any other scientific factor.[3] Now, at the beginning

of the twenty-first century, the World Health Organization's fifteenth model list of essential medicines (2007) contains 325 individual drugs, including 12 antiretroviral medicines for the prevention and treatment of HIV/AIDS. The economic impact of drug research and prescription drugs is now overwhelming. According to an NCPA report in 2000, France spends 1.6 per cent of its gross domestic product (GDP) on drugs, Britain 1.1 per cent and Japan 1.5 per cent. In 2007 health-care in the USA was 6.6 per cent of GDP (which was $2.2 trillion) and retail spending on prescription drugs reached a record $227.5 billion.[4]

In academia, receptors are now the *raison d'être* for most research in pharmacology and pharmaceutical sciences, and understanding their role in signalling in the nervous system lies at the very heart of neuroscience. However, the receptor is also a keystone concept in other scientific and medical disciplines, including biochemistry, immunology, chemotherapy, molecular biology, physiology and toxicology. An indication of this significance can be seen in journal citations. From a simple search of the National Library of Medicine's Gateway database, for articles published between 1960 and 1969, there were 5107 journal citations with *receptor* or *receptors* in the keywords, title or abstract; between 1990 and 1999, there were more than 314,000 such journal citations. The distinctive position that the drug receptor concept holds in pharmacology leads medical historians as well as pharmacologists on a quest to learn more about the historical development of this core idea of modern medical science. This book deals with the history of the receptor concept and aims to present the development of this idea in its contemporary context.

Research methods and approaches

The receptor concept was born in the last decade of the nineteenth century. To write a history of scientific ideas from a cultural perspective was the inspiration behind this book, but it was also a problem. Venturing into the area of the 'history of ideas' seems to be a somewhat outdated approach. The 'history of great ideas' is one of the classic topics of the traditional 'Whiggish' history of science and medicine, and descriptions of revolutionary concepts and innovations fill the pages of many of the older monographs in the field. This old style of writing tends to portray the history of medical concepts as a progress-oriented success story ending in the era of modern scientific medicine. This style of writing on the history of ideas is still prevalent as representatives of various medical disciplines publish flattering accounts of their respective specialities.[5]

Practising physicians or medical researchers often link the rise and fall of ideas purely to the thoughts and concepts of the scientists concerned.

By contrast, the challenge we faced was that of describing and analysing the history of the drug receptor concept in its cultural context. This follows the more recent trend in studies of scientific and medical concepts where they are related to research institutes and hospitals, methods, experimental settings, epistemic objects, researchers, patients and the public in general – to mention only a few. Meanwhile, there are many publications that describe well-defined research cultures, each with a specific style of thought, experiment and research. Often, these publications focus on isolated aspects, for example, the involvement of politics in scientific decision-making, research and gender problems, research and therapeutic reasoning, or the relation between medical experimentation and social and medical institutions.[6] For good reasons, most recent edited volumes on the history of medical innovations try to label their contributions according to the topics mentioned above. They aim to demonstrate different perspectives on the history of innovations and to describe and analyse specific factors which influence the research process.[7] There remain two problems in relation to this procedure. First, there is the danger of losing sight of the fact that a multitude of quite different aspects may be involved in the development of a particular idea. Second, the history of concepts or ideas is not itself treated as a separate research topic – perhaps because of the fear of slipping back into the old Whiggish formula or because science studies venture into this area from other starting points.[8]

This book delivers a history of the receptor concept as an idea in its cultural contexts. Here we apply the word 'culture' in its broadest sense, as the actions and rituals of human beings in ordering their world.[9] Our refusal to narrow the term 'culture' enabled us to identify many different factors and aspects which have moulded the story of the receptor concept. This way, our approach resembles that of the historian of chemistry, Jack B. Morrell, who 'suggested a way in which social history of science could be written which did not downgrade science as cognition, which bypassed the sterile dichotomy between internal and external history of science, and which avoided any form of naïve reductionism'.[10] Morrell's work focuses on research schools and scientific networks, including factors such as recruitment, training, the careers of scientists, and the power of directors of scientific institutions. The cognitive side of science and its social and cultural practice are very much interrelated, and this is what we want to show in this book when writing about the history of the receptor concept.[11]

As the historian Peter Moraw has pointed out, researchers do not leave their social connections in the cloakroom when entering the lab.[12] Indeed, the birth of the receptor concept was the outcome of circumstances in the lives of its two founding fathers, the physiologist John Newport Langley (1852–1925) and the immunologist and bacteriologist Paul Ehrlich (1854–1915). Scientific debates, career issues, religious faith and politics invaded the lab, or at least influenced the direction of research. Langley and Ehrlich independently invented the concept of receptors, but it was an idea not solely developed on the basis of scientific knowledge. Especially in the case of Ehrlich, it involved fantasies partly born from novels. The final concept was so flexible that it was resistant to falsification, and Ehrlich promoted 'receptors' like a new product. The concept was then discussed internationally, but within a context of competing theories and a scientific community of pharmacologists that adhered to rigid nineteenth-century traditions. The concept was also confronted with contemporary trends in medicine that favoured the improvement of therapy rather than lengthy programmes of basic research. Langley, Ehrlich and other protagonists of the idea died early or were relative outsiders in pharmacology. Receptor research was disrupted by the two world wars, the concept was hard to operationalize with current techniques and was challenged by competing research strands. But it finally made its breakthrough.

Although the receptor concept eventually proved to be an effective tool in pharmacology, ours is not a success story. It is a story of originality, but also one of chance, of lucky and unlucky coincidences, of ups and downs. Even today, the concept is debated. With hindsight it is less astonishing that it took 60 years for the concept to be accepted, than that it was finally accepted at all. The fate of the concept depended on social networks in pharmacology and medicine. It was developed and applied by single researchers and their collaborators (research groups) in Germany (Ehrlich), the UK (Langley, Clark, Black), and the USA (Ahlquist, Moran). But between around 1905 and 1950 it was also denied and rejected by networks of leading pharmacologists/physiologists in Germany, the UK and the USA.

The starting points for our analysis are the actors and their works. The papers of the various researchers, published in contemporary journals, enabled us to trace the development of individual theories. But the papers are complemented by the examination of textbooks that reflected the state of the art and the standing of the receptor concept in contemporary pharmacology. Moreover, archival materials as well as the oral history method have been used to complement the published sources and to contextualize the receptor research. This helped us analyse the biographies of the main actors and the circumstances of their lives. Such material also helped us consider the contemporary fashions in medicine, the zeitgeist, together with the scientific networks that influenced the fate of the receptor concept. Archival sources were examined in Berlin, Freiburg, Graz, Cambridge, Edinburgh, London, New York and Augusta; they ranged from specific biographical material to material on institutions.[13] Oral history was restricted to specific cases, where gaps in knowledge and explanation still existed and could not be closed with printed and unprinted materials. This is especially true for the research on Raymond P. Ahlquist (1914–83), who played an important role in promoting the receptor concept after 1945. Because of all the problems linked with this method,[14] we see oral history only as an auxiliary tool, the results of which have to be cross-checked with textual material whenever possible.[15]

Using these sources as a base, it was possible to write the history of a scientific concept informed by approaches of social constructivism. We focused on pharmacology as a field that was predominantly inspired by the new idea of receptors, and the history of this special discipline of medicine formed the background of our work.

The background: theories of drug action before 1900

Theorizing about the ways in which drugs and poisons act on the human body is as old as our Western medical tradition. Such ideas have been closely linked with contemporary understandings of how the body functioned and of the nature of disease.

In Greek antiquity, Hippocrates (*c.* 460–*c.* 370 BC) and his followers defined physical health as the balance of four cardinal humours (fluids) of the body: 'blood', 'phlegm', 'yellow bile' and 'black bile'. Each of these four humours was characterized by a pair of primary qualities. Blood was described as hot and wet, yellow bile as hot and dry, phlegm as cold and wet, and black bile as cold and dry. Disease was interpreted as an imbalance between the four humours. Treatment therefore aimed at restoring the equilibrium, usually by regulating the patients' lifestyle, that is, exercise and rest, sexual activity, clothing, housing and, in particular, diet. If, for example, a patient suffering from a fever was considered to be overly 'hot and dry', the Hippocratic healer advised eating foods that were 'cold and moist', such as barley gruel or seafood. If an excess of a humour was diagnosed, the opening of a vein (phlebotomy) was thought to remove this excessive matter through bloodletting.

Drugs were often understood as a means to evacuate excessive or harmful humours and were thus classified according to the way they expelled matter: as emetics (vomiting), purgatives (laxatives), diaphoretics (sweating), diuretics (increasing the excretion of urine), expectorants (facilitating the coughing up of mucus), cholagogues (supporting the excretion of bile) and emmenagogues (bringing forth menstruation). Primary qualities were also ascribed to drugs, and treatment followed the principle of *contraria contrariis curantur* (opposites are cured by opposites). A phlegmatic patient, for example, who suffered from an abundance of cold and wet humours, required treatment with a heating and drying drug such as thyme.

The remedies of Hippocratic medicine were mostly made from plants, but some were also made from substances of animal origin, for example cuttlefish eggs, deer horn, blister beetle or castoreum (from the beaver). A number of mineral substances such as common salt, soda, alum, sandarach (disulfide of arsenic), antimony and some copper salts were also employed. The knowledge of a drug's effects appears to have been derived from therapeutic experience and (largely oral) tradition.[16]

In Roman times, the imperial physician Galen of Pergamon (AD 129–*c.* 200) developed a more elaborate pharmacological system on the basis of the Hippocratic principles. This attributed the four primary qualities to drugs and, in addition, stipulated certain degrees of efficacy, from zero up to four or five. Water, for example, was described as cooling at a degree of zero and rose-water as first-degree cooling.[17] Opium, according to Galen, was cooling in the fourth degree and the flesh of a viper moderately heating and strongly drying.[18] The rationale of Galenic drug treatment still followed the Hippocratic principle of *contraria contrariis* in applying remedies that produced effects opposite to those of the disease. For example, a 'hot' fever required 'cooling' drugs. Drugs with opposite qualities were also combined in the hope of producing compound remedies that had intermediate degrees of efficacy and that would be both safe for the patient and appropriate for many conditions. The most famous of these were universal remedies or panaceas, such as 'theriac', which included both viper flesh and opium, in addition to many other ingredients.[19]

Like the Hippocratic healers, Galen also advocated therapy through evacuating overabundant humours. He theorized that each drug attracted the humour that was proper to it, in the same way that the body attracted its appropriate nutrients or a lodestone attracted pieces of iron. The cholagogue drug scammony, for instance, was said to attract yellow bile from jaundiced patients, and safflower and *Cnidian* berry to draw phlegm from the body. There was even some speculation in Epicurean thought (with which Galen did not agree) that substances that attracted each other became entangled or interlocked through minute hook-like extremities.[20]

In ancient Greco-Roman culture, though not strictly in Hippocratic medicine, such 'rational' attempts at pharmacology and pharmacotherapy often went hand-in-hand with religious and magical forms of healing. As the cause of disease might lie in both environmental conditions and divine wrath, praying to the gods, sleeping in the temple of Asklepios (the God of Healing), wearing amulets and performing diverse magical practices were believed to complement physical forms of treatment.[21]

During the Middle Ages, the Hippocratic humoral theory and Galen's doctrine of drug effects were authoritative guides for treatment. Reflecting the Christian worldview, however, writers emphasized that the success of a therapy always lay ultimately in the hands of God. This combination of religious and naturalistic or rational approaches to healing continued into the Renaissance (and beyond), regardless of the wider revolutionary changes that were brought about by the Reformation, the Copernican system and the discovery of the New World. Paracelsus (1493/94–1541), the most outspoken medical revolutionary of the period, despised Galen's teachings, but advocated the doctrine of 'signatures', according to which God had given certain signs to plants to allow human beings to recognize their healing properties. Thus a yellow plant such as saffron indicated its healing power in cases of jaundice, or the leaves of *Pulmonaria*, which display lung-shaped markings, hinted at their usefulness in treating respiratory diseases. With Paracelsus, alchemy became a major influence on therapy. He characterized substances according to three chemical principles: the inflammable 'sulphur', the volatile 'mercury', and 'salt', that is, the residuum after an alchemical procedure. Under Paracelsus' influence, so-called 'chemical' medicines of mineral origin, containing iron, lead, copper, sulphur, antimony, arsenic or mercury, gained prominence in pharmacotherapy. They had, in part, been used already in Hippocratic medicine, but their use by Paracelsus and his followers corresponded to a new chemical interpretation of bodily processes.[22]

In the seventeenth century, iatrochemistry (medical chemistry) developed from these Paracelsian ideas as a specific orientation of learned medicine. Diseases were understood as the product of inner 'fermentations' and as an imbalance between acidity and alkalinity in the body. Depending on the diagnosed acid or alkaline character of a condition, an alkaline or acid medicine was prescribed in order to neutralize the excess. Other theories of drug action reflected a vitalistic conception of disease. According to the Flemish physician and philosopher Jean Baptiste van Helmont (1579–1644), disease was an expression of an organ's disturbed vital principle, the *archaeus*. Remedies had a specific 'taste', the *sapor specificus*, which enabled the *archaeus* to recognize them as beneficial.[23]

Under the influence of René Descartes (1596–1650) and his followers, mechanistic interpretations of bodily functions as well as of the actions of drugs developed as an alternative to those vitalistic and chemical speculations. Iatromechanism, which flourished in the late seventeenth and eighteenth centuries, attempted to apply Robert Boyle's (1627–91) corpuscular understanding of chemical processes and Isaac Newton's (1642–1727) notion of gravity to physiological and pharmacological phenomena. Poisonous mercury sublimate, for example, was believed to cause inflammation and gangrene of the stomach and the guts through minuscule 'fiery spikes', and metallic mercury was thought to dissolve 'coagulations' and to act as a purgative through the weight and motion of its rotund particles.[24] During this period the concept of a 'specific' remedy, that is, of a drug that healed a specific disease through some hidden property or unexplainable power, was launched by those who were critical of chemical or mechanistic speculation. The English physician Thomas Sydenham (1624–89) famously praised the (quinine-containing) Peruvian bark, a drug first brought to Europe by Jesuit missionaries in the 1630s, as a true 'specific' against intermittent fever (that is, malaria). Diseases were supposed to be classifiable in the same way as plants in botany, and specific remedies against specific kinds of disease were supposed to be identified on an empirical basis.[25]

Mechanistic, chemical, vitalistic and empiricist approaches to the understanding of drug effects continued to compete with each other throughout the eighteenth century and beyond. Distinctive systems of medicine that developed in this period had characteristic predilections in pharmacotherapy, for example, the rival systems of Friedrich Hoffmann (1660–1742) and Georg Ernst Stahl (1659–1734), who for a time both taught medicine at the University of Halle in Germany. Hoffmann and his students propagated mechanistic interpretations of drug action; they proposed that opium, as a hypnotic and analgesic drug, worked by thinning the blood, which distended the arteries of the brain or made blood serum seep out of the vessels, pressing on and obstructing the nervous fibres. For Stahl, by contrast, opium dazed and stupefied the soul, which not only guided physiological movements, but was also the force behind the body's salutary expulsion of harmful matter. Stahlians were therefore very sceptical about the therapeutic value of opium, which (as they saw it) paralysed the 'healing force of nature', whereas Hoffmann's followers had no objections against using this powerful drug in treatment, if it was applied with due caution.[26]

Alongside the various scientific interpretations, religious views continued to be relevant. Hoffmann, shortly before his death, wrote a work on natural theology or 'physicotheology', in which he praised God's providence in having put salutary mineral springs in the earth, and in letting certain plants grow specifically in those places and countries where they were most needed for the treatment of certain endemic diseases. Other writers used the example of useful medicinal plants to demonstrate, in the teleological fashion of natural theology, 'God's power, wisdom and benevolence' towards mankind.[27]

From the late seventeenth century, chemical tests, *in vitro* experiments on blood, animal experimentation and human trials were employed to explore the mode of action of drugs and poisons. Often such experiments were used to support a particular point of pharmacological theory. Some trials, however, were performed to decide on the general question of whether poisons acted 'by sympathy', for example by stimulation of nerve endings in the stomach wall which propagated the effect through the body, or by way of absorption into the blood and distribution through the circulation. In the early nineteenth century the theory of drug effect via absorption became dominant, especially through the animal experiments of the Paris physiologist and physician François Magendie (1783–1855).[28]

At this time new pharmacological theories and treatments had become fashionable. Brunonianism, named after the Scottish physician John Brown (1735–88), claimed that illness was characterized by either a lack of bodily excitement (asthenic diseases) or over-excitement (sthenic diseases). To restore a balance, a mixture of alcohol and opium (laudanum) was used as a stimulant for asthenic conditions, and a vegetable diet or bloodletting was recommended for sthenic diseases.[29] Towards the end of the eighteenth century, the German physician Samuel Hahnemann (1755–1843) introduced the system of homoeopathy. Breaking with the Galenic principle of treatment by *contraries*, Hahnemann taught that substances producing symptoms *similar* to those of the disease should be used in therapy. This principle of *similia similibus curentur* (treat like with like) together with the principle of 'potentiation' of a drug's effect by way of extreme dilution (to transfer the drug's power to the solvent) alienated homoeopathic physicians from their more conventional colleagues.[30] Despite this variety of new theories of drug action, medical practice largely continued to follow Galenic principles until well into the nineteenth century.

As was the case in other areas of medicine, pharmacology was, for a period, influenced by Romanticism, especially in the German-speaking countries. Following Newton's work on attraction and repulsion around 1700, chemists throughout Europe had become interested in studying the chemical affinities between substances. By the beginning of the nineteenth century, this theme had become part of a wider culture. In 1809 Johann Wolfgang von Goethe (1749–1832) used the concept of '*Elective Affinities*' ('*Wahlverwandtschaften*') in his novel of the same name to describe and understand human relationships. Unsurprisingly, the concept of affinities also featured in Romantic pharmacological writings, especially as it had similarities with Galen's ideas about the selective attractive powers of drugs. Karl Friedrich Heinrich Burdach (1776–1847) used these terms to explain the difference in pharmacological efficacy between substances, and Friedrich Sobernheim (1803–46) postulated 'specific elective affinities' ('*spezifische Wahlverwandtschaften*') between certain drugs and body parts. In Sobernheim's view, strychnine had a specific affinity to the spinal cord, digitalis and tobacco to the nerves of the heart, alcohol to the brain, mercury to the salivary glands, ergot to the nerves of the uterus, and sulphur to the skin.[31]

Yet in the same period pharmacology experienced a decisive empirical turn, first in France with the experimental work of Magendie, and then in German universities, where pharmacology was first institutionalized as a laboratory-based medical discipline. Magendie's programme for physiological research was simply to record the phenomena of life obtained through experiments and to distrust any higher, vitalistic 'principles'. It was hoped that the observed phenomena would be reducible, eventually, to physical and chemical laws. This experimentalism and reductionism applied also to Magendie's pharmacological work. Benefiting from the contemporary work of pharmacist-chemists, from 1813 onwards he examined, in animals and healthy and ill human beings, the effects of various alkaloids that had been isolated from plant sources, such as morphine (from opium), emetine (from ipecacuanha roots), quinine (from Peruvian bark), and strychnine (from *nux vomica*), as well as pure chemical substances such as prussic acid and iodine. This emphasis on studying the effects of 'pure' substances rather than the traditional compound remedies marked an important new departure in the history of pharmacology.[32] Thus, depending on new developments in chemistry and particularly physiology, Magendie's experimental approach was adopted by others, especially his most famous student, the physiologist

Claude Bernard (1813–78), as a model to study the effect of drugs and poisons as well as physiological processes in general. Bernard, for example, showed in animal experiments that the point of attack for the paralysing arrow-poison curare was where the endings of the motor nerves met the muscle fibres.[33]

This type of pharmacology was initially deeply embedded in physiological research as the subject became institutionalized as a laboratory discipline. At the University of Dorpat, Estonia, in 1847, the German professor Rudolf Buchheim (1820–79) created the first laboratory for experimental pharmacology. It was initially self-funded and used by his doctoral students, one of whom, Oswald Schmiedeberg (1838–1921), went on to become, in 1872, professor of pharmacology at the new German Reich University of Strasbourg, established after the Franco-Prussian war of 1870/71. Alsace had been annexed to Germany, and Strasbourg was generously funded by the state to make it the German 'model university'. Schmiedeberg was thus able to found a large institute of pharmacology, which soon attracted postgraduates and visiting researchers from many countries and set an example for the establishment of similar institutes at other German universities.[34]

Both the Buchheim-Schmiedeberg school of experimental pharmacology and the French line of drug research within experimental physiology avoided theorizing about the nature of drug action. While the modes and sites of action of many substances, as well as their metabolism in the animal body, were studied extensively, broader pharmacological theories were not formulated. As Schmiedeberg wrote in 1867 (when still an assistant to Buchheim in Dorpat) with regard to theories about the action of chloroform, there were too many premature theories. Better, then, 'with all the means of physics and chemistry [and] on the basis of physiology [to] establish new, indubitable facts through experimentation and observation'.[35] This emphasis on observable 'facts' reflected the general positivism of the natural sciences of the nineteenth century, and the influence of the Buchheim-Schmiedeberg school on the study of drugs continued well into the twentieth century (see Chapter 3 below).

It was unclear, however, how certain drugs selectively affected particular tissues or organs. A strand of pharmacological theory developed from the notion that specific relations between the chemical structure of a substance and its effects in the body might be identifiable. Magendie himself had hinted at this and one of his former students, the English physician James Blake (1814–93), demonstrated in the 1840s that inorganic compounds with the same macroscopic crystalline structure produced similar physiological effects when infused intravenously. By the 1860s chemistry had sufficiently developed to enable other researchers to show structure-activity relations for organic substances as well. In London, Benjamin Ward Richardson (1828–96), later known for his contributions to public health, studied the action of various amyl compounds in the frog and found that slight modifications in the chemical composition of a compound led to small variations in its effects on the animals. He thus suggested that the chemical law of substitution might have its counterpart in a 'physiological law of substitution'. In Edinburgh, a similar line of investigation was followed by the pharmacologist Thomas Richard Fraser (1841–1920) and the chemist Alexander Crum Brown (1838–1922), who compared the physiological action of salts and substitution products of various alkaloids, including strychnine, morphine, codeine, nicotine and atropine. They showed, for example, that whatever the 'normal' effect of the alkaloid, a change in one of the nitrogen atoms (from tertiary to quarternary form), invariably produced a curare-like paralysing action.[36] However, this kind of detailed structural study was still rare; chemistry was not routinely used in medical drug research until the twentieth century.

By the early twentieth century, a scientific controversy had developed over whether pharmacological action depended directly on a substance's chemical structure or rather upon its physical properties, and therefore only indirectly upon its chemical constitution. Advocates of a chemical theory of drug action faced opposition from adherents of the so-called 'physical theory',[37] a controversy which formed the backdrop to the development of the receptor theory of drug action. Another important general question was how substances acted upon animal and human *cells*. Rudolf Virchow's (1821–1902) classic work on *Cellular Pathology*, published in 1858, established the view that disease could be understood as an expression of morphological changes and disturbed functioning in cells.[38] Microscopic investigations using dyes that selectively coloured certain types of cells and cellular structures became important. From the 1890s attempts were also made to study the selective action of poisons on cells, for example by the Leipzig pharmacologist Rudolf Boehm (1844–1926), and subsequently by one of his pupils, Walther Straub (1874–1944).[39] The old notion of 'affinities' between certain drugs and parts of the body gained new significance through this research. This was the immediate background of the development of the receptor concept, and it is therefore in this period, between the 1870s and 1890s, that the narrative of this book begins.

Outline of the book

Chapters 1 and 2 analyse the origins and early development of the receptor concept by the Berlin immunologist and Nobel laureate Paul Ehrlich and the Cambridge physiologist John Newport Langley. Between 1878 and 1905, these two scientists approached the idea from quite different fields of research outside of pharmacology. Ehrlich was chiefly interested in bacterial toxins and antitoxins; he came to the idea of cell receptors via the development of his immunological *side-chain theory*, which explained the mechanism by which toxins are bound by immune cells and by which antitoxins are produced in the body. Langley was mainly interested in the physiology of the autonomic nervous system, and he examined the effect of drugs and poisons on nerve endings and muscles. He postulated the *'receptive substance'* as the site of action for nicotine- and atropine-like drugs. Langley and Ehrlich constructed two different receptor theories which came together only after 1905. The first two chapters of this book show how the development of the receptor concept by Ehrlich and Langley was an outcome not only of their intellect and originality, but also of their personal decisions and career paths, and of several different issues then debated in medical research.

The receptor idea was not, however, immediately accepted in pharmacology. Chapters 3 and 4 examine the debate about the receptor idea and competing theories of drug binding and the transmitter concept. Chapter 3 argues that resistance to the idea of receptors arose after 1905 because there were several competing theories of drug action, and, with the technologies available, it was difficult to test these ideas. Furthermore, the main representatives of academic pharmacology, such as Walther Straub in Germany and Arthur Robertson Cushny (1866–1926) in England, were strong advocates of a theory of drug action centred on physical, rather than chemical, explanations. Chapter 4 is devoted to the history of the transmitter concept; it explains why influential representatives of physiological and pharmacological research, including Sir Henry Hallett Dale (1875–1968) and Otto Loewi (1873–1961), were less interested in receptors than in transmitter substances in the nervous system. Chapters 3 and 4 together show that the empiricist experimental style developed by Rudolf Buchheim and Oswald Schmiedeberg and their reluctance to build broad pharmacological theories continued to dominate the scientific community of pharmacologists.

Chapter 5 deals with the elaboration of the receptor concept in the late 1920s and early 1930s when it regained prominence through the work of the British pharmacologist Alfred Joseph Clark (1885–1941). He was initially open-minded about the 'physical' and 'chemical' (receptor) theories of drug action and adopted a quantitative approach to better understand the effect of drugs on cells. He determined that certain *minute* quantities of drugs could influence cells only through specific areas of the cell, and concluded that receptors must be responsible for the transmission of drug effects (*receptor occupancy theory*). Clark's successor in the Edinburgh Chair of Materia Medica, John Henry Gaddum (1900–65), went on to develop the concept of competitive antagonism of drugs for receptors. Subsequently, Clark's receptor occupancy theory was further refined and modified in the 1950s by R.P. Stephenson (1925–2004) (also in Edinburgh) and by the Dutch pharmacologist E.J. Ariëns (1908–2002). But throughout this period, receptors remained hypothetical entities.

Chapters 6 and 7 cover the consolidation of the receptor concept in pharmacology. Although the receptor theory became quite widely accepted by the 1950s, it was not the central focus of pharmacological research. As Chapter 6 makes clear, even the important studies of the American pharmacologist Raymond P. Ahlquist (1914–83) did not immediately change this situation. In 1948, Ahlquist divided the receptor for the neurotransmitter adrenalin into alpha- and beta-receptors, a distinction that was crucial to the impact and later pharmacological exploitation of receptors. The evidence for the existence of different sub-types of receptors answered many questions that had arisen from the multiple and often opposing effects of a single drug in the body. Ahlquist, however, was a relative 'outsider', his personality and his cautious way of presenting his results, set against the dominant background of the neurotransmitter research, were not conducive to a wider dissemination and adoption of the receptor concept. After more research on chemical mediation of drug effects, and the modifications of Clark's theory by Ariëns and Stephenson, it became easier to apply Ahlquist's refinement in clinical pharmacology. A critical breakthrough for the receptor concept arrived with selective receptor-blocking drugs: in 1965, Sir James Black introduced the first therapeutically useful ß-blocking agent, propranolol. For his work on receptor-subtype-selective drugs he received the Nobel Prize in Physiology or Medicine in 1988.

Chapter 7 describes how the consolidation of the concept was achieved with the isolation and purification of receptor proteins and their visualization in the late 1970s and 1980s. Since then a multitude of receptors and receptor families have been described and their mechanisms of signal transduction delineated. This last chapter has been written from the perspective of a neuropharmacologist (Robert Halliwell) who is active in this field of research. Given our historical closeness to this period, the narrative provided here is somewhat more technical and involved than in the previous chapters. However, the non-specialist reader will take from it an appreciation of how important the receptor concept has become to biomedical research and how receptors were eventually transformed from more or less useful hypothetical constructs into material objects of scientific inquiry.

We want to show that the history of the development and impact of the receptor idea on pharmacological problems by no means represents a simple success story. On the contrary, we demonstrate that the receptor idea, although developed in the last years of the nineteenth century, went through periods of acceptance and rejection before critical pharmaco-therapeutic breakthroughs occurred in the 1960s. While certain episodes of the 'receptor story' have been told in articles and book chapters by other authors, most notably John Parascandola, Joseph Robinson and Arthur Silverstein,[40] our book is the first to give a comprehensive and detailed examination of the origins and development of the receptor concept and a historical explanation for its slow acceptance in pharmacology. The history of the idea of receptors serves as an example of how research programmes and medical disciplines are shaped by the combination of intellectual, biographical and social factors.

1
Paul Ehrlich and his Receptor Concept

This chapter focuses on the Berlin bacteriologist and immunologist Paul Ehrlich (1854–1915). It deals with the emergence of his receptor concept between 1878 and approximately 1905, by which time the concept was largely developed. As Ehrlich was one of the pioneers of the receptor idea, research on Ehrlich and his receptors needs to solve a basic problem: are receptors 'objective' facts of nature and the 'discovery' of the receptor somehow 'inevitable', or did Ehrlich 'construct' the receptors?[1] We will draw on the second alternative, which corresponds with the sociological constructivist interpretation of science: that discoveries do not neatly correspond to objective entities in nature; they depend decisively on the cultural setting, for example, the social position of the researcher and the local scientific system.[2] Ehrlich's discovery of the receptor concept emerged from both his social and scientific backgrounds. Although the roots of his ideas can be traced back to early stages of his career, although his ideas appeared in a logical order, and although he was driven by a leitmotif throughout his academic life, it was far from clear that he would develop the receptor concept. And the receptor concept, as developed by Ehrlich, was nothing more than a hypothetical option, on the one hand vulnerable to falsification and on the other hand a prospective tool for explaining parts of the metabolism of animals and human beings. In this chapter we will concentrate on the origin of the receptor idea. Its birth and early character were rooted in problems that made the introduction of the idea in post-1945 pharmacology a surprising and unforeseeable event.

First, Ehrlich's basic scientific ideas and the traditional views on his academic life will be discussed. Second, we will look more closely into Ehrlich's career in relation to his work on receptors. Finally, the findings will be summarized and evaluated.[3]

The roots of Ehrlich's research and the construction of the receptor

If we focus only on Ehrlich's ideas, the introduction of his receptor concept appears to be consistently planned and realized. As a medical student at the universities of Breslau, Strasbourg, Freiburg and Leipzig, he was already concerned with the staining of histological specimens. This means that Ehrlich was confronted with new research strands in medicine even in the early periods of his education. Histology and the staining of tissues and cells were new methods within the pathological tradition that had started to develop at the beginning of the nineteenth century. The interest in the formation of organic structures, formerly invisible, was fuelled by the introduction of the achromatic compound microscope around 1830, enabling detailed investigation of the microstructure of tissues and cells, with quite a good resolution. The microscope became an important tool in anatomy as well as pathology and from about 1830 Johannes Müller (1801–58), anatomist and physiologist at the newly founded University of Berlin, trained numerous students in using the new tool for research purposes. One of Müller's students, Rudolf Virchow (1821–1902), propagated the routine application of the microscope in medicine to investigate pathologically changed tissues and cells. From about 1850 the method of staining that had been developed by Joseph von Gerlach (1820–68) in 1847 was used, starting with carmine red. In 1870, at the Carl Zeiss Factory in Jena, Ernst Abbe (1840–1905) developed the oil immersion lens with a much better resolution, thus also enabling the investigation of microorganisms.[4]

With such a rapid development of histopathology, Ehrlich started his staining experiments under his first teacher, the anatomist Wilhelm von Waldeyer-Hartz (1836–1921). Ehrlich followed him from Breslau to Strasbourg, and it was Waldeyer-Hartz who stimulated Ehrlich's fascination with histology, which did not diminish when he returned to Breslau two years later. He was then given a post at the institute of the pathologist Julius Cohnheim (1839–84), one of the few German pathologists who used animal experimentation in addition to pathological anatomy.[5] Cohnheim introduced Ehrlich to research into pathological functions, which involved close contact with clinicians, in contrast to the study of pathological morphology. Cohnheim also influenced many British medical students and contributed in this way to the introduction of 'clinical pathology' to Britain.[6] Under the influence of Cohnheim, Ehrlich followed the path of applying laboratory findings to clinical practice, becoming a pioneer of clinical pathology in Germany.

In addition to pathophysiology, Ehrlich's knowledge of histopathology and staining was bolstered through contact with his cousin Karl Weigert (1845–1904). In Breslau, Ehrlich also made contact with Robert Koch (1843–1910), who demonstrated the finding of the anthrax bacillus in the institute of the plant physiologist Ferdinand Cohn (1828–98). These important influences propelled Ehrlich towards success in his studies at Freiburg University. He stained, among others, 'plasma cells',[7] so-named by Waldeyer-Hartz. Later, in his 1878 dissertation based on this work, Ehrlich distinguished the so-called 'mast cells'[8] from these 'plasma cells'.

Histology and histopathology, staining, the idea of investigating disease processes in a pragmatic way, the use of the microscope and staining to diagnose infectious diseases with the aim of combating them – these were all facets of Ehrlich's student education, impacting on his future academic career. In this context, Ehrlich's own core scientific ideas can be summarized in two points: first, he made a claim for the routine application of the method of staining in medicine in general and in histology in particular, and second, he supported the theory that the staining process relied on a chemical reaction between dye-stuff and cell.[9] These two ideas formed the basis or leitmotif of Ehrlich's side-chain and later receptor theory.

There are four different steps that can be described as important events in the development of Ehrlich's receptor concept. The first step is his dissertation on the 'theory and practice of histological staining', which he wrote in 1878. In this publication, he was already writing about 'a definite chemical character of the cell'[10] which was necessary for its reaction with a dye. The second step was his habilitation thesis, which he wrote in 1885 on the 'oxygen-need of the organism'. Dyes served to measure animal organ oxygen usage and he developed the theory that the protoplasm of the cell had 'side-chains' to bind oxygen.[11] The third step was his publication in 1897 on the evaluation of diphtheria serum. Working on immunological problems, Ehrlich returned to the 'side-chains' and developed his chemical 'side-chain theory'; certain 'side-chains' of the cell were able to bind certain toxins. Because these occupied side-chains would then become unable to fulfil their physiological functions, the cell would overcompensate by producing a lot of additional side-chains. These side-chains would be released into the bloodstream, where they acted as antibodies or antitoxins.[12] The fourth and last step was the introduction of the term 'receptor' in 1900. Ehrlich introduced the term 'receptor' simply as a designation for the side-chain's function.[13] In 1908, Ehrlich was awarded the Nobel Prize for his work in immunology.

In the following years, the receptor concept also served to explain effects in the therapeutic application of substances. Ehrlich first thought receptors would bind toxins and nutritive substances only. A shift in his thinking enabled him to apply his concept to drug binding. Ehrlich's former idea was that many drugs could easily be extracted from tissues and therefore it would seem that they could not be bound firmly to the cell. Therefore they could not provoke the production of side-chains.[14] Ehrlich's theory allowed – as a future perspective – drug binding to cells only indirectly, as he proposed that certain chemical bodies or groups (*Körperklassen*) with specific binding capacities to specific organs could be used as 'vehicles' (*Lastwagen*) to carry artificial substances to the site of effect.[15] No earlier than 1907 – with reference to his own work on the effect of dyes on trypanosomes and John Newport Langley's notion of 'receptive substances' – Ehrlich accepted the binding of drugs to receptors.[16] For this purpose there would be specific 'chemoreceptors'. The therapeutic drug needed to have a greater affinity to the chemoreceptors of the invading microorganisms, for example, the trypanosomes, than to the chemoreceptors of the host's body. The concept of 'specific affinity' was applied to the pharmacological realm. Based on early nineteenth-century investigations of the relationship between chemical composition and the physiological action of certain drugs,[17] Ehrlich was now able to explain all phenomena connected to the effect of a chemotherapeutic drug, for example, drug resistance, with the help of the receptor concept. The receptors became the theoretical basis for his subsequent work with the dye 'trypan red' and with the arsenic compound 'atoxyl' to combat trypanosome infections, resulting in the discovery of Salvarsan, the first chemotherapeutic substance for the treatment of syphilis, in 1910.[18] Consequently Ehrlich became one of the most prominent figures of twentieth-century medicine.[19]

The development and realization of Ehrlich's theories was soon revealed to be a three-stage success story: staining – immunology – chemotherapy.[20] The hypothetical character of the side-chain and receptor theory was soon acknowledged, but its development within the frame of Ehrlich's life and career was described as a process of maturing or even as the successive appearance of plans already secretly developed. This interpretation did not remain unchallenged, but it shaped the historiography of Ehrlich's work.[21] Even in Ehrlich's own time,[22] the side-chain or receptor theory found its place in his celebrated biography. In 1919, one of his students, Leonor Michaelis, was impressed by the re-discovered inaugural dissertation of the master. It would be 'of great value for the recognition of Ehrlich's scientific development'. According to Michaelis, the dissertation foreshadowed the whole of Ehrlich's work. It seemed as if Ehrlich knew right from the beginning about the institutional frame of his work he wanted to set up and above all that his lab work would lead to enriching medicine with a new theory of life.[23]

Even today, Ehrlich's dissertation is sometimes regarded as the master plan for a successful research programme.[24] Recent papers on his work, which do not venture extensively into Ehrlich's biography and the circumstances of his life, explicitly or implicitly defend the narrative of a continuous path towards the receptors, based on a strand of complicated and theory-laden laboratory work. Anthony S. Travis explains the development of the side-chain and receptor idea as an outcome of Ehrlich's continuous and progressive work on dyes. For him, considering the 'social frameworks' meant examining Ehrlich's collaboration with the chemical industry. The main point to note is the support of private sponsors in promoting Ehrlich's theoretical immunological research. Timothy Lenoir's interpretation reflects this thinking, positioning Ehrlich's research on the side-chains within the triangle of science, politics and industry. Also, as far as Lenoir is concerned, there is no doubt that Ehrlich's worksite is the lab and that his instruments are his Petri dishes, his test tubes and his research animals. In 1999, Arthur M. Silverstein drew an even more radical conclusion from his analysis of Ehrlich's 'receptor immunology'. According to him, the receptor idea had been pursued by Ehrlich for over 20 years, from his dissertation up to 1898. In his monograph on Ehrlich's receptor concept, Silverstein basically followed the same argument. He mainly focused on Ehrlich's immunological work on diphtheria and the smooth development of the receptor concept as a sort of natural outcome from venturing into the realms of biochemical lab work in late nineteenth-century Germany.[25] The studies of John Parascandola and Ronald Jasensky aptly describe the ups and downs of the early receptor theory, but do not explore its social, historical and cultural context.[26] These unidirectional and occasionally simplistic approaches raise some questions since, as we saw above, Ehrlich got much more input in his student days. Apart from histopathology and staining, he also experienced broader pathophysiological thinking, which was not restricted to lab research on functional changes in diseased tissues and cells. Cohnheim also envisaged clinical applications and Koch hoped to contribute to combating the big threat of the nineteenth century – infectious diseases.

In order to discover what Ehrlich really wanted at each stage of his life, we have to take a closer look at his life. As we will see, the examination of the conceptual development in the context of Ehrlich's biography reveals a much more complex story, with many factors conditioning the emergence of his key idea. When analysing Ehrlich's development of the receptor theory within the framework of his socialization and academic career, one can detect five phases, which will be dealt with chronologically.

Paul Ehrlich's scientific career and the emergence of his receptor concept

(A)   Ehrlich as a clinician, 1878–88

In 1909, a few years before his death, Ehrlich told a friend what he considered to be the most important aspects of his scientific work. He pointed out that he had 'always' had the greatest interest in therapeutics and that the combination of his chemical and therapeutic interests could explain his entire scientific career.[27] This comment, made with hindsight, reveals that there was something more going on than basic research in the laboratory. And 'always' means that his interests in therapeutics cannot be restricted to the years after 1905, when he was involved in the tiring and sometimes unpleasant discussions about Salvarsan.[28] In fact, very early in his career, Ehrlich was involved in therapeutic work on the wards.

In 1878, Ehrlich was appointed as a physician at the Medical Clinic of the Charité Hospital in Berlin under the well-known professor of internal medicine, Theodor Frerichs (1819–85).[29] From the time of his predecessor Johann Lukas Schönlein (1793–1864), the clinic had stood for a solid foundation of therapy built on the carefully developed diagnostic skills of the physicians and on the unification of practice and theory in medicine. Schönlein promoted percussion and auscultation of the patients on their sickbeds in combination with chemical and microscopical analysis of body parts and fluids. Frerichs, whose efforts were based very much on contemporary experimental physiology, wanted to integrate laboratory research and laboratory diagnostics into clinical work on the wards. He supported Ehrlich's staining experiments[30] because he too combined his clinical work with his work at the laboratory bench in subsequent years. Ehrlich examined tissue specimens from the postmortem room to gain knowledge about the causes of patients' deaths, and staining methods helped to explain the pathological functions of the morphological structures. A good example is Ehrlich's study of glycogen in healthy and diabetic human subjects.[31]

Ehrlich also examined the body fluids of living patients. For example, he studied the pleuritic exsudates (effusions into the pleural cavity) of women in childbirth, wherein the application of different staining methods on microbes allowed him to identify different infections, and to give diagnostic and prognostic advice.[32] In the years under Frerichs, Ehrlich also performed the first biopsies of the liver on humans.[33] Finally, he described the so-called Diazo-reaction (*Diazo-Reaktion*) in 1883, a special urine test to detect bilirubin with the help of dyes – used for patients thought to have heavy infections.[34]

Only occasionally did Ehrlich publish single clinical case histories with only marginal relations to his scientific work.[35] Most of Ehrlich's papers written in these clinical years are concerned with the application of dyes, especially when conducting animal experiments. In 1882, Ehrlich examined fluorescent dyes on the eye of the rabbit, hoping to find a method to diagnose human eye diseases; the effectiveness of the substance as well as possible side effects were of interest.[36] His experiments on blood were outstanding: again following on from animal experimentation, different dyes were applied and tested on the blood of healthy and ill persons. Different kinds of white and red blood cells were explored, and the experimental work was then applied to the diagnosis and therapy of both blood diseases and infectious diseases. Thus, Ehrlich became a pioneer in modern haematology,[37] while in bacteriology he collaborated with Koch. After 1882, when Koch had demonstrated the newly found tubercle bacillus, Ehrlich improved Koch's staining method.[38]

Ehrlich used dyes not only to solve problems of diagnosis but as potential therapeutic agents. In 1886, the substance Thallin was applied to rabbits and shortly thereafter to patients suffering from typhoid fever. The best dosage to combat fever symptoms was estimated, particularly in the climax of the disease. Ehrlich was able to give advice on the clinical usage of this substance.[39]

The 'side-chain' theory was the outcome of one specific study – his habilitation thesis (for his teaching licence) of 1885, which was dedicated to Frerichs. Using countless animal experiments, Ehrlich tried to estimate the 'oxygen-need of the organism' (*Sauerstoff-Bedürfniss des Organismus*) and his dyes served to indicate the different oxygen affinities of the various organs. Following the infusion of dyes the animals were killed after certain intervals of time, and dissected. The colouring of the tissues indicated the metabolic activities of the organs and tissues and enabled a certain classification. According to Ehrlich the 'protoplasm' of cells had side-chains that could bind oxygen; the resulting complexes could be burned in the protoplasm and transformed into energy. His study presented a theoretical basis for his staining methods, but the side-chain as predecessor of the receptor played no important role in his discussion of oxygen utilization. Moreover, Ehrlich focused on the organs, the character of the protoplasm, and that of the surrounding paraplasm of the cell (*Paraplasma*, the more unspecific areas of the cell, *die mehr indifferenten Territorien des Zellenleibes*).[40] The side-chains were mentioned only six times in Ehrlich's 69-page habilitation thesis (Himmelweit edition of the collected works) and never again during the period of his work under Frerichs.[41]

Ehrlich's major aim was to achieve clinically applicable results. He could influence the work on the wards by connecting different fields: staining methods, animal experimentation, clinical work and human therapeutic experiments – occupations that filled his entire working time. He achieved his aims without any deeper insights into the exact way in which substances bind to cells, which explains why there was no discussion of the side-chains in the years after the publication of Ehrlich's habilitation thesis. In 1891 Ehrlich was still calling himself a 'clinician',[42] and he had acquired his teaching licence in the field of 'practical and clinical medicine' (*praktische Medicin und ärztliche Klinik*).[43] And in 1898, when, together with a colleague, he published the results of his research on the blood cells, he pointed out the importance of clinical examination. When describing the places of origin of the white blood cells, Ehrlich wrote that it would be '... hard to avoid errors if one confines oneself exclusively to animal experiments without supplementing these by clinical experience ... Not the anatomist, not the physiologist, but only the clinician is in a position to discuss these problems.'[44] It was not necessary to apply his idea of side-chains (published in the meantime) to his haematological results, although they were concerned with metabolic and microbiological problems, for example, the effect of bacterial poisons on white blood cells.[45] Ehrlich was mostly devoted to laboratory work, but this made sense for him only in connection with practical medicine: he tried to improve diagnostic techniques and tools to make them usable even for the untrained physician walking the wards. He was not concerned about developing his ideas about side-chains. The two areas mentioned later by Ehrlich as most important – chemistry and its application to biological problems and therapy – did not require any detailed knowledge about receptors.

Private as well as professional matters developed well. In 1883 Ehrlich married. In 1884, his parents moved from Strehlen in Upper Silesia to Berlin,[46] and in the same year, Ehrlich became a titular 'professor'.[47]

(B) The change of emphasis, 1889–95

Ehrlich's situation changed rapidly after Frerich's death in 1885. The Medical Clinic of the Charité Hospital was now renamed the 'First

Medical Clinic' and handed over to Ernst von Leyden (1832–1910), who had been professor of internal medicine in Berlin and head of the so-called 'Preparatory' (*Propädeutische*) Clinic of the Charité Hospital since 1876. The resulting vacancy at the Preparatory Clinic was filled by Carl Gerhardt (1833–1902), who had been professor of internal medicine in Würzburg. The Preparatory Clinic itself was renamed 'Second Medical Clinic'.[48]

Ehrlich now worked under Gerhardt in the Second Medical Clinic of the Charité in Berlin. Gerhardt supported the traditions of the Medical Clinic, but there was a clear shift of interest in the direction of meticulously performed clinical investigations of the patients on the wards. The attached laboratories were still seen as important parts of the clinic, but moved a bit into the background. Above all, Gerhardt controlled the clinical work of his assistants, who were under pressure to meet his high demands.[49] Gerhardt focused on the further development of clinical diagnostics and on the organization of empirical studies on patients, and he integrated Ehrlich into the daily clinical routine. There is enough evidence to show that this caused severe problems for Ehrlich, who remained a devoted laboratory worker. He caught tuberculosis in 1888, resigned and went to Egypt for recovery. To leave his position in Berlin was a difficult break in his career, particularly because, as a Jew, Ehrlich could not obtain a full professorship nor employment at a state institute.[50]

In 1889, Ehrlich returned to Berlin.[51] He was unemployed and could set up a small laboratory only with the financial help of his father-in-law. Ehrlich no longer had any patients and had to rely solely on his dyes and animal experimentation. He turned to immunological work. The latter was inspired by Robert Koch and contemporary ideas about anti-bacterial treatments, and particularly by Koch's assistant Emil von Behring (1854–1917) who, in 1890, together with his co-worker Shibasaburo Kitasato (1852–1931), had discovered the phenomenon of antitoxins in diphtheria and tetanus.[52] During this period, Ehrlich was successful in immunizing mice against the plant poisons, ricin and abrin; he investigated the suggested hereditary transmission of immunity and its transmission via breastfeeding; and he obtained results on the basic processes of active and passive immunization.[53]

In 1890, only one year later, the period of private science ended when Robert Koch offered his former helper a post as clinical supervisor for scientific studies on tuberculosis at the City Hospital Berlin-Moabit. Now Ehrlich was able to take up clinical experimentation again, as he had done during his time with Frerichs, and he again combined it with animal experimentation and histological investigations in the laboratory. Koch gave him a small laboratory and a few assistants.[54] Together with a colleague, he explored the best tolerated dosage of the tuberculin serum and combined therapeutic human experimentation with histological sputum examinations.[55] Ehrlich again worked with dyes in analysing the analgesic effect of methylene blue. He had gained some clues, via his staining experiments, about a certain affinity of the substance for nerve cells, especially the axon cylinders of sensory and motor nerves. These trials on methylene blue were performed on prisoners of the Moabit Royal Prison (*Königliche Strafanstalt Moabit*) and the connected Observation Unit for Insane Criminals (*Beobachtungsanstalt für geisteskranke Verbrecher*). The dye methylene blue was shown to have an effect on certain types of pain, above all, on migraine.[56] Furthermore, methylene blue was tested as a therapeutic agent against malaria.[57]

In 1891, Robert Koch offered Ehrlich a laboratory of his own in the newly founded Institute for Infectious Diseases (*Institut für Infektions-krankheiten*). This was an important career step for Ehrlich, since Koch's institute was one 'of the most prestigious' of the bacteriological institutes founded in Europe around this time. Solely devoted to research and freed from teaching, researchers like Ehrlich could develop their scientific talents extensively.[58] And it was a good opportunity for Ehrlich to cooperate with the bacteriological working group under Koch, which included Behring, Richard Pfeiffer (1858–1945) and August Wassermann (1866–1925). Ehrlich's interest in immunology grew enormously. The ability of the body to form substances (that is, antibodies) to combat specific microbes seemed to be useful for developing new treatments: the therapeutic sera. From 1891 Ehrlich worked chiefly on human immunology, but while the investigation of dyes and clinical tests moved into the background, they continued to play a role in his daily work. Diphtheria serum, for example, was tested on children.[59]

(C) Ehrlich as a theorizer, 1895–1905

But Ehrlich remained in a dependent position, and the direction of his work turned again as a result of pressure from his 'working group' under Koch and from a sponsor outside the institute. A request from Behring meant a return for Ehrlich to the investigation of theoretical problems of immunity. Behring and the Hoechst Company (*Farbwerke Hoechst*, near Frankfurt/Main) had been having difficulties with the production of the new therapeutic diphtheria serum. Up to 1894 it had not been possible to produce it in reliable concentrations. As in the case of tuberculin, the challenge was to standardize the effective dosage. Behring asked Ehrlich for help. Considering the keen competition in Koch's institute, one can assume that it was only 'the direct request of Behring, urged on by an impatient Koch and a cost-conscious Hoechst Company, that would allow Ehrlich to venture into an area to which his institute colleague had full priority claim'.[60] Behring agreed with Ehrlich that he should examine the exact quantitative relations between diphtheria toxin and antitoxin and develop a method to standardize the application of the therapeutic serum.[61]

The research on the diphtheria serum helped to put Ehrlich's scientific career onto a socially secure basis. But in Koch's institute, Ehrlich never received an official post because of his Jewish faith.[62] When in 1895 the 'control station' for therapeutic sera (*Controllstation für Heilsera*) was opened in the institute, Ehrlich became only the deputy head of the department, whereas the responsibility for its work was handed over to Koch's assistants August Wassermann and Hermann Kossel.[63] Eventually, however, the influence of the powerful ministry councillor in the Prussian Ministry of Science and Education (*Ministerialrat im preußischen Kultusministerium*), Friedrich Althoff (1839–1908), saved Ehrlich from the problems and difficulties with his provisional post. Ehrlich and Althoff knew each other very well and were also in private contact.[64] It was Althoff who promoted Ehrlich's career further and who organized the main institutional framework of his academic life. Ehrlich was grateful, as he believed the academic community had judged him as 'unusable'.[65] A safe social position was important and had to be secured. This was so important that even Ehrlich's wife wrote letters to Althoff. She was still trying to secure an independent position for her husband in 1903, when the main decisions on Ehrlich's institutional setting had already been made.[66] The challenge of the diphtheria serum brought Ehrlich an independent position. At the instigation of Althoff, he became, in 1896, head of the new Institute for Serum Research and Serum Testing (*Institut für Serumforschung und Serumprüfung*) in Steglitz in the suburbs of Berlin. Its main purpose was the testing of sera, but it also enabled Ehrlich to focus on diphtheria research in the laboratory.[67]

These events and the new configuration led to Ehrlich's second, more theoretical phase of work, and they instigated the creation of his 'side-chain' and 'receptor' theory. Although relevant as an explanatory tool for his research, the 'side-chains' had been mentioned only in two of his papers between 1885 (the year of the development of his theory) and 1897.[68] But in his classic study on 'the assay of the activity of diphtheria-curative serum and its theoretical basis' (1897)[69] Ehrlich defined the side-chains as a part of an immunological system. He was trying to find a standard dosage for the application of diphtheria antitoxin, and used guinea pigs to test a vast amount of diphtheria toxins of different origin. This enabled him to find two threshold concentrations. The first one was a completely neutralized solution of toxins, which caused no sign of disease when applied to a guinea pig. The second one described the quantity of toxin that killed a 250-gram animal within four days. The difference between the first (neutral) and the final (lethal) solution was called the 'single lethal dose' (*einfache letale Dosis*).[70]

Ehrlich's results were far from encouraging as the concentrations of the solutions of the different diphtheria toxins varied markedly. Moreover, the solutions were not stable, but lost their toxicity after a certain period of storage, although the number of antibody-binding units did not change. This meant that the toxic effect did not correspond to the capacity of the toxin to bind to antitoxin. The only explanation for this phenomenon was that the toxins themselves had undergone some changes.[71] Forced to explain these results, Ehrlich came back to his 'side-chain theory'. The toxin was thought to consist of two parts: a poisonous component, the so-called 'toxophore group' (*toxophore Gruppe*), and a component which enabled the binding to the antitoxin, the so-called 'haptophore group' (*haptophore Gruppe*). According to Ehrlich the toxophore group was not as stable as the haptophore group, and in consequence, the toxophore groups successively dissolved; thus toxins emerged which were able to bind antitoxins but no longer had a toxic effect. Ehrlich called these poisons 'toxoids' (*Toxoide*). Here was an explanation for the instability of many antitoxin-solutions and a key to working out programmes to maintain the standard of diphtheria antitoxin samples.

The toxoids were also a key to understanding immunological processes. Through their binding capacity to a side-chain of the cell, these toxoids were able to induce the production of antibodies: the latter were an overcompensated production of side-chains, which were released into the bloodstream. This meant that chemical processes of specific binding were combined with biological processes of regeneration.[72] Ehrlich explained the binding mechanism with the analogy that the biochemist Emil Fischer (1852–1919) had used to describe the effect of enzymes on substrates when he wrote about the 'key-lock' mechanism. There was a strong overlap in the work of Ehrlich and Fischer because the latter was interested in the synthesis and constitution of bodily substances. The term 'enzyme' had been coined in 1878 by Wilhelm Kühne (1837–1900), physiologist at Heidelberg University. It described molecules consisting of proteins, whose responsibility is the transformation of substances into new ones based on building an enzyme-substance-complex. Fischer had elaborated Kühne's ideas with the concept of protein chains as the basis of the enzyme structure and action. In 1902, Fischer received the Nobel Prize for Chemistry for his work on sugars and purines. That Ehrlich used the 'key-lock' analogy is not surprising, as he loved to use metaphors to describe and explain difficult immunological processes and functions. Furthermore there was no theoretical hindrance to Ehrlich using and accepting the enzyme idea since it correlated with his basic assumption of transformation through the binding of his toxin-antitoxin (or antibody) complex. The blow to Ehrlich's ideas of antibody formation did not come until the work of Landsteiner in the second decade of the twentieth century: Landsteiner was able to gain credit for his idea that the antibody did not originate in cell material but could be triggered by new substances which had entered or invaded the organism (see also sub-section E below). Nevertheless, the 'key-lock' analogy in particular proved to be of basic importance for explanatory models that were developed in biochemistry over the course of the twentieth century.[73]

With his 1897 paper, Ehrlich developed his hitherto vague and provisional idea of side-chains into a 'side-chain theory'. From now on, this theory would be the basis for his immunological investigations. Ehrlich's work became theory-oriented, and immunological studies dominated his research in subsequent years.[74]

(D)   From the 'side-chains' to the receptors

In the following years Ehrlich's side-chain theory became more and more intricate as its details were studied experimentally. Between 1897 and 1905, these difficult experiments were carried out with an extensive use of animals of different species and of poisonous substances. Before 1899, the experiments were performed in the Steglitz Institute,[75] thereafter in the Institute for Experimental Therapy (*Institut für Experimentelle Therapie*) in Frankfurt/Main. The latter was set up for Ehrlich by the Prussian state, again with the support of Althoff.[76] Ehrlich was able to organize and to coordinate experimental studies using numerous assistants. His right-hand man and coordinator for the laboratory investigations on the side-chain theory was the bacteriologist Julius Morgenroth (1871–1924), who had been Ehrlich's assistant in Steglitz since 1897.[77] In 1906, after the research on the side-chain or receptor theory had come to a notable conclusion, Morgenroth became head of the department of bacteriology at the pathological institute of the Charité Hospital in Berlin, walking in the footsteps of Ehrlich and performing research on immunology. In

1919, Morgenroth became head of the department of chemotherapy of Robert Koch's Institute for Infectious Diseases.[78]

Morgenroth was very important as a collaborator and he disseminated Ehrlich's concept at a later stage in his career, contributing most prolifically to the development of the receptor concept compared to other assistants. Most importantly for us, however, Ehrlich himself continued to be the *spiritus rector* of the whole project. Ehrlich urged his assistants to perform more and more animal experiments to consolidate his theory. His laboratory notes show this, for instance, when he demanded: 'Please also show me the Pyrodin-animals', or simply 'Where is the ape?'[79] Accordingly, the 'side-chain theory' expanded conceptually. The 'toxoids', those poisons only capable of cell-binding, were supplemented by the 'toxons' (*Toxone*) only one year later, in 1898.[80] The toxons were also ineffective poisons, but in contrast to the toxoids, they were synthesized and released by the microbe itself and had not lost their toxophore group.

Besides analysing poisons, Ehrlich concentrated on the mechanisms and processes of the side-chain theory itself. The animal experiments on haemolysis, that is, the solvent action of antibodies on red blood corpuscles, seemed to be helpful. This could be compared with antibodies attacking bacilli. Some studies on haemolysis, published together with Morgenroth, enabled Ehrlich to improve the side-chain theory decisively. The results of these studies led to the investigation of double-binding side-chains, which – according to Ehrlich and Morgenroth – were also responsible for the process of nutrition.[81] A second binding site on the side-chain enabled specific molecules to anchor and to induce the processes of haemolysis and of nutrition respectively. In this way Ehrlich opened up the path from the more or less narrow immunological terrain to the understanding of the general metabolism of the human organism. The side-chain theory appeared to be a good explanatory tool to uncover the deepest secrets of biological chemistry. In 1899, Ehrlich postulated a countless number of side-chains that would adapt to the 'constantly changing chemistry' of the body. This chemistry would be influenced by race, sex, nutrition, energy, secretion and other factors, and so there were continuous changes taking place in the blood serum.[82]

Finally, in 1900, Ehrlich and Morgenroth, in their third paper on haemolysis, introduced the term 'receptor': 'For the sake of brevity, that combining group of the protoplasmic molecule to which the introduced group is anchored will hereafter be termed receptor.'[83] The side-chains *as such* had played a minor role in Ehrlich's research until this point of his career. He had concentrated on the relations between the side-chains on the one hand and the poisons and bodily substances on the other, but not so much on the character of the side-chains per se. The titles of the published papers illustrate this. Even the third and the fourth papers on haemolysins are chiefly about the classification of the different kinds of 'complement' (Ehrlich's name for a molecule anchoring at the second binding site of the side-chain and enabling the process of haemolysis).[84]

The introduction of the term 'receptor' was not merely a new term for an old idea.[85] After having elaborated the meaning of the 'side-chains' with the creation of a 'side-chain theory', it was a logical next step to investigate the specific nature of side-chains. The new term stood for the central position of the side-chains in Ehrlich's new research strand. The 'receptors' and their specificity for certain poisons soon played an important role in Ehrlich's immunological theory.[86] In 1901, Ehrlich and Morgenroth, when deepening their research on this general topic, made some remarks on the 'peculiarities of the receptor apparatus'. They mentioned that there would be a vast number of receptors attached to the red blood cells and that these receptors could bind to diverse immune bodies and haemotoxins.[87] As in the case of the antibodies or intermediate bodies, the structure of the receptors was analysed and classified. As complex receptors with two haptophore groups, the inter-mediate bodies (antibodies regularly belonging to the immune system) were now called 'amboceptors' (*Amboceptoren*).[88] This view of receptors characterized them as flexible entities, first of all attached to the cell and binding toxins or immune bodies, but then also doing the same in the bloodstream.[89] However, Ehrlich soon conceded that there were receptors that, even in the case of a successive overproduction of the cell, would *not* be released into the flowing blood. Instead they remained at the surface of the cell. Ehrlich called them 'sessile receptors' (*sessile Receptoren*).[90] But they were only one subspecies. Some receptors were common to different animal species, and one could find them in many organs. A receptor could have many complement binding groups, therewith mutating to a 'triceptor' (*Triceptor*) or 'quadriceptor' (*Quadriceptor*).[91] The step towards postulating the existence of a 'polyceptor' (*Polyceptor*) able to bind many substances was eventually taken in 1905.[92] One year before that, in 1904, Ehrlich had already divided the receptors according to their ability to bind substances, distinguishing first, second and third order receptors.[93] All those with two or more binding sites belonged to the receptors of third order. Blood plasma was filled with a vast amount of receptors released into the bloodstream, which Ehrlich called 'haptins' (*Haptine*). These haptins included all those substances that at the time were not yet identified.[94] Finally, he approached what he called the 'pluralistic point of view' (*plurimistischer Standpunkt*).

This was the assumption that there were a whole range of different complements, anti-complements, receptors and many other substances.[95] Because receptors played a key role in metabolism, research on receptors was not only aimed at classifying different substances, but also at more or less indirectly shaping the character and activity of the receptors themselves. And, because they played a key role in immunology and in combating microbes,[96] as well as in the physiology of nutrition and human metabolism in general, Ehrlich predicted a great impact of the receptor concept on clinical medicine. He believed that his studies on receptors would 'open a new meaningful direction of biological research'.[97]

(E)   The fight for the receptors and lost alternatives

Ehrlich's experimental methods were refined again and again, for example, with the application of new chemicals. He received confirmation and support from colleagues, but he could not provide direct evidence for his theory, as the receptors were not visible.[98] The experimental setting and his exact and thorough methods enabled Ehrlich to identify the microcosm of substances only indirectly, from reactions of the blood sera. This continued to be problematic since Ehrlich himself provided speculative drawings of the receptors. On the one hand, Ehrlich knew that his receptor theory had to remain a theory for the present. But, on the other hand, because it was based on experimental evidence, he thought it legitimate to publish the results.[99] The receptor theory shares certain characteristics with the atomic theory of new theoretical physics emerging at the end of the nineteenth century. Although unconfirmed, it was at least seen as helpful in explaining certain phenomena of nature. Consequently, Ehrlich's theory was thrown a lifeline with the help of confirmatory ideas and thoughts that were developed by researchers in the early twentieth century.[100]

In 1901, Ehrlich maintained that the side-chain theory had 'passed the test perfectly'.[101] It could show and explain the immunological response of organisms to infectious diseases and also show and explain basic metabolic procedures in animals and man. The appearance of scientific phenomena that were inconsistent with his theory was explained by Ehrlich with the highly intricate conditions of the animal organism. The inconsistencies would serve to deepen the theory and to promote its success.[102] This meant that a theory should not be dismissed as soon as any contradiction occurs, but should be carefully rethought. Ehrlich saw himself as a pioneer of a new medicine of the future, and he compared his 'side-chain theory' with Rudolf Virchow's 'cellular pathology', which had also not been immediately successful.[103]

The uncertain basis of Ehrlich's theory provoked criticism from the outset. The critics – and Ehrlich's way of dealing with them – served only to lead him deeper into his work on the receptor concept. Because of their argumentation, Ehrlich attributed to many of his critics what he called the 'unitarian view' (*unitarischer Standpunkt*). As far as he was concerned, the 'unitarians' rejected the multitude of substances that were necessary for the metabolism of the body, holding, for example, that there would be only one complement that would bind to receptors to induce agglutination of the blood. This stood in sharp contrast to his own 'pluralistic conception' (*plurimistische Anschauung*), which was grounded in the idea of many receptors and a variety of different options to bind substances, depending on the kind of receptors and on the condition of the cells and tissues. Based on their own experiments, the 'unitarians' attacked specific points of the side-chain theory and thereby questioned Ehrlich's microcosm of immunological substances.[104] Ehrlich identified one opposing group of scientists at the Pasteur Institute in Paris. This bacteriological institute had been founded in 1888 as 'a living memorial' to Pasteur and stood in keen competition with Koch's Institute for Infectious Diseases. Although both institutions profited from each other in terms of research methods and organization, in an age of imperialism they were symbols of national scientific success in France and Germany respectively.[105] This nationalism clearly overshadowed communications between Ehrlich and the Pasteur Institute, but as these tended to concentrate on scientific matters there were few real problems. The opposing group at the Pasteur Institute consisted of the French immunologist and bacteriologist Emile Roux (1853–1933), deputy director from 1895 and director of the Pasteur Institute from 1904; his Belgian colleague and co-worker Jules Bordet (1870–1961); and the Russian immunologist and bacteriologist Elias Metchnikoff (1845–1916) who had worked at the institute since 1888. The dispute between these three researchers on the one side and Ehrlich on the other started in about 1897 and continued for about ten years.[106]

Most important was the conflict with Jules Bordet. In 1900, Bordet questioned the direct binding of side-chains and complements as well as the variety of immune bodies that had been postulated by Ehrlich.[107] Such attacks annoyed Ehrlich considerably and his response reveals his methods of dealing with critics of his theory. Ehrlich instructed his assistants to repeat Bordet's experiments; under the heading 'against Bordet' (*gegen Bordet*) he gave them directions on pieces of paper. The results of the experiments did not lead to a revision of the side-chain theory, but to its vindication, and the newly obtained immunological knowledge was integrated into Ehrlich's flexible, pluralistic theory.[108] In this way Ehrlich and his workers not only defended the side-chain theory against its detractors, but also consolidated it. Ehrlich had a restless correspondence with colleagues. In his letters, he reported immediately on his newest findings and dissected the experiments of his opponents, describing them as failed criticism of the side-chain theory. For example, Ehrlich reported to a colleague that Metchnikoff was the 'real spiritus rector' of the attacks from Paris, having soft-soaped Roux with his 'breathtaking personality'. It was 'such a shame that such an experimenter and such a clear head as Roux got so deeply into mysticism and into the Russian fog'.[109]

Ehrlich had friendships with many of his scientific opponents, but when it came to defending his theory, he launched a massive campaign against any critic, not only openly, but also behind the scenes. Ehrlich complained about seemingly disparaging remarks in the secondary literature. Once he tried to urge the editor of the journal *Deutsche Medizinische Wochenschrift* to stop the printing of critical remarks by one of his opponents, because his own views 'had gained full acceptance already'.[110] Ehrlich sorted his colleagues into friends and enemies of his theory. In October 1902, he wrote to William Henry Welch (1850–1934): 'I was most delighted to recognize you as one of the warmest friends of the theory, but even more so that you could achieve such new and fundamental insights with its help.' In contrast, he wrote to a pharmacologist in Halle (Germany), 'that every impartial person reading the literature has to count you as an absolute opponent'.[111] Again, Ehrlich related the remark to 'the theory', which occupied him more and more in the years after 1897. Comments made by Ehrlich's critics give us further insight into the obsessive nature of his involvement in the receptor concept in this period.

Ehrlich's fierce style of rejecting criticism was also apparent in his controversy with the physical chemist Svante Arrhenius (1859–1927) and his student Thorvald Madsen (1870–1957), which started in 1903. Ehrlich had to defend his biological point of view against a physico-chemical interpretation of antitoxin-toxin binding. Arrhenius and Madsen applied basic chemical laws to processes of life, which according to Ehrlich, could not be expressed in such rigid formulas. Worst of all, Arrhenius adopted Bordet's unitarian point of view on the haemolysis of blood.[112] Ehrlich again carried out experiments and started a letter-writing campaign against Arrhenius.[113] In a letter to Althoff, presumably written in

1904, Ehrlich noted that Arrhenius would be 'pushed to the wall'.[114] This was of course, an exaggeration. Physico-chemical explanations gained ground in the following decades, but this outcome was hard to foresee during the first years of the twentieth century, when there was no generally accepted drug binding theory and different scientists worked to promote their own approaches.

Even more serious was Ehrlich's dispute with Max (von) Gruber (1853–1927), professor of hygiene in Vienna from 1887 to 1902 and in Munich from 1902 to 1923. Although giving him credit for new findings in the field of immunology, Gruber attacked Ehrlich's substance-binding theories of specific toxins and immune bodies in a very polemic way, as purely speculative with a nearly total lack of evidence. For Gruber, many immunological phenomena were still unexplained and applying Ehrlich's theory to the solution of specific problems represented no more than a logical game. Gruber launched attacks against Ehrlich's approach in several papers on immunology, published between 1901 and 1903.[115]

Again, Ehrlich made tremendous efforts to explain to his academic friends the weakness of Gruber's criticism. Ehrlich entirely rejected the latter's comments as 'stupid' (*blödsinnig*) and treated them as a '*quantité négligeable*'. In return, Gruber pointed out that 'my only criticism is that in the course of his theorizing he [Ehrlich] permits too much fantasy and too little criticism'.[116] Gruber, the 'gifted polemicist'[117] evidently hit Ehrlich in a vulnerable spot: all the inventions of new terms and substances could not really explain the formation of antibodies. For Gruber, 'only the conditions that accompany the processes of life are accessible to our research'.[118] Ehrlich, despite his adverse attitude towards Gruber's remarks, published two papers dedicated to his opponent in 1903.[119]

The strongest blow to Ehrlich's side-chain and receptor theory came from a group of immunochemists who attacked the chemical specificity of the concept on unitarian grounds. The most important representative was the Viennese pathologist and immunologist Karl Landsteiner (1868–1943), a student of Gruber in 1896, who had joined the front-line of Ehrlich's opponents. Landsteiner subsequently developed the colloid theory of immunology: that reactions were influenced by the chemical constitution of substances, but above all by physical phenomena such as solubility and temperature. Like Gruber, Landsteiner attacked Ehrlich's 'uneconomical' *plurimism* (or pluralism), as loaded with too many uncertainties. From 1903 onwards, Landsteiner invaded the field of immunology with his theory,[120] by 1912 'the colloid theory had superseded Ehrlich's, although in the practice of the serum institutes the old assumption of clear-cut, one-to-one specificity was essentially unchanged'.[121]

All these critics did not lead Ehrlich to rethink his 'pluralistic' view. On the contrary, as many printed and unprinted sources show, he became more and more obsessed with his theory in the years after 1900. Ehrlich had only ever had a meagre interest in culture and politics, and there are no indications that he recognized his involvement in the state-oriented and nationalistic bacteriological research of his teacher Koch.[122] Ehrlich talked mostly about his work and drew speculative sketches of the receptors on note-pads, letters, postcards and even on the floor or tablecloths.[123] And as we know from a recent contribution form Cambrosio et al., the images themselves fuelled the criticism of opponents, as they were an attempt by Ehrlich to illustrate invisible structures whose material existence was not evident but debated at this time. For Bordet, Ehrlich's sketches were responsible for the success of the side-chain theory; it was based not on facts but on an imagery that was wrongly interpreted as the picture of real material life.[124] Remarkably enough, Ehrlich's efforts to consolidate the receptor theory were supported in the realm of literary fiction. One of the few things Ehrlich did in his leisure time was to read detective stories, especially those of Sir Arthur Conan Doyle (1859–1930). Evidently, these also served as a way of contemplating his work, as he filled the margins with comments and formulas.[125] The ideas of his teacher Robert Koch flooded back into his mind through the reading of poetry. Conan Doyle (himself an ophthalmologist by training) was an admirer of Koch and did not hesitate to travel to Berlin in 1890 to investigate the bacteriologist's tuberculin cure. Doyle's first wife suffered for many years with tuberculosis, but his fascination with the idea of preventing the invasion of the empire by microbes also came to shape the work of his fictional hero, Sherlock Holmes, who concentrated explicitly on detecting the 'invisible' and subversive threads of life. Sherlock Holmes searched for traces of a crime that could only be detected after the closest scrutiny or with the use of tools; in his search for microbes, Koch was in the same position. Doyle saw both Holmes and Koch as 'imperial knights' who saved their empires.[126]

For Ehrlich, experimental findings and the sketches of invisible immunological entities[127] together ensured the credibility of the side-chain and receptor theory in the same way that visual representation (early photography) and animal experimentation had demonstrated Robert Koch's postulation of living pathogenic microbes.[128] Ehrlich explained that his sketches were 'merely a pictorial method' and therefore did not correspond with reality.[129] But such protestations fade in comparison to the extensive usage Ehrlich made of his pictures to persuade his contemporaries of the truthfulness of his theory. Criticism allowed Ehrlich to reconstruct and enlarge his receptor concept, since his strategy focused on further investigating his opponents' viewpoints and integrating newly obtained knowledge into his theory, saying that his colleagues' counter-arguments would finally confirm the receptor concept. Ehrlich compared himself to a victorious chess player who cannot finish the game because his defeated opponents do not want to give up.[130]

Based on the shift from practical clinical work to entirely theoretical work, Ehrlich's intense concentration on side-chains and receptors decreased only slowly until 1905, in parallel with the increase of his cancer research and finally – from about 1906 onwards – with his work on chemotherapy. Here again, he came back to therapeutic medicine. At this point one has to consider that the impact of Landsteiner's immunological theory on Ehrlich's second shift of interests is questionable. Still, in 1906, Ehrlich criticized colloidal chemistry.[131] He thought his own theory to be completed, and he now used it as a tool to examine other fields of medical research.[132]

Despite Ehrlich's enthusiasm about receptors, there are indications in his correspondence that he was not entirely satisfied with his situation. Even in the last Berlin years at the institute in Steglitz, that is, after 1898, Ehrlich was faced with the problem that, as head of a theoretical institute, he had no patients.[133] The reason for his wish to have patients was the often repeated intention that 'after having worked entirely on questions of serum therapy for such a long time, I want to turn a bit to my old pet subject again, which is histological and biological staining'; or, expressed in another way: 'After the long period of immunological magic I am now getting around to focus again a little bit on the dyes as my old pet subject.'[134] Although predominantly publishing on the side-chains between 1897 and 1905 (28 papers), he did not stop publishing on dye-stuffs and their clinical application (6 papers) or on purely chemical problems (2 papers).[135] Even in his immunological studies he did not hesitate to draw parallels with staining processes. Ehrlich was not very keen to work as a physician again, but he wanted to carry out clinical testing of potential drugs. In the following period, he pre-tested dyes on animals, examined the side effects, and thereafter sent the dyes to clinicians with the request to perform therapeutic human experiments with them. Ehrlich ordered the dye-stuffs from pharmaceutical companies, then he arranged for them to be packed into capsules before handing them over to physicians. This was meant to simplify the application to the patients.[136] The bottleneck in this system was the patients, and in the following years Ehrlich constantly begged the clinicians to perform therapeutic experiments. His troubles forced him to come back to those he knew well, for example his old friend Albert Neisser (1855–1916), the professor of dermatology in Breslau. In 1898 Ehrlich wanted Neisser to administer some of his dyes to patients for therapeutic reasons. These dyes had already been tested on rabbits and hares and had been well tolerated. Ehrlich recommended a slow increase in the dosage 'first of all in cases of headaches, vague rheumatic pains, gonorrhoea and cystitis'.[137] Neisser was willing to test the dye 'brilliant blue' on patients. In November 1898 Ehrlich became pushy and urged Neisser to speed up the trials: 'After all, however, it is not so difficult – considering your great experience with these things – to find the approximate dosage *bene tolerata*.'[138] Neisser, who was involved in a scandal over performing human experiments on prostitutes and children without information or consent, tested several dyes for Ehrlich and discussed the chosen patients with him.[139] Other colleagues were less helpful. Ehrlich repeatedly asked them to test his substances. Even if the clinicians were prepared, in principal, to undertake the experiments, Ehrlich had to press them forcefully to do the test and to report the results. The system did not work well and from June 1899 at the latest Ehrlich showed signs of frustration. He remarked to a colleague that 'all these gentlemen undertake the staining therapy more or less to do me a favour but not out of deep conviction'.[140]

In the end, dye testing appears to have been a futile attempt to restore the old Charité conditions, where laboratory work and animal experimentation could be linked with clinical expertise. Indeed, Ehrlich's favourite style of work was so well known that in 1899 the administration of the city of Frankfurt feared that the patients of the City Hospital would be 'used for experimental purposes'.[141] But what endured of his first years in Frankfurt were the academic laboratory studies on receptors, based on animal trials and test-tube experiments. Although basically satisfied with his independent position and his working conditions,[142] Ehrlich also outlined the drawbacks when he described his work on the receptors in 1901: 'Because I am myself not in a position to perform such investigations on a large number of patients, I thought it to be my duty to clarify my point of view and in this way to lay the basis of the work in a field whose importance for pathology and therapy presumably will be fully acknowledged only after many years.'[143] In 1905, this argument was essentially repeated when he explained that he had done his work and that 'more new and successful work' could only be done by 'specialists, who have the necessary *clinical* and pathological material'.[144]

This did not mean that Ehrlich wanted to become a full clinician. In the same year, presumably because of the responsibilities towards his institute, he rejected the call to become director of the First Medical Clinic in Vienna.[145] He wanted to be attached to clinical facilities as a laboratory worker. Ehrlich continued to be dependent on distant collaboration with physicians in respect to drug trials – for example with his teacher Robert Koch.[146] And finally, it was due, in part, to a revival of the old idea of therapeutic application of dyes, the application of trypan red to trypanosomes, that Ehrlich could successfully develop Salvarsan. In combination with these old ideas, the theoretical work on the side-chains and receptors had its practical impact.[147]

Paul Ehrlich, pharmacology and the receptors

The literature published so far on Ehrlich describes the clear, unidirectional development of Ehrlich as a laboratory worker with a clear-cut aim which waited to be achieved. In contrast, our findings show that while he did indeed have a theme there was no well-developed master plan leading him straight to the receptors. At the beginning of his time in Berlin, under the clinician Theodor Frerichs, Ehrlich advocated a concept that did not rely on the intricate construction of the side-chain theory as a predecessor of the receptor concept. Ehrlich's provisional ideas on the theoretical basis of his work did not play an important role in his main aim in these years, namely to achieve practical therapeutic results. These results could be achieved by combining, in a pragmatic way, his favourite branch, chemical laboratory work, with animal experimentation and therapeutic human experimentation, and there was no need to deepen the knowledge on the side-chains. The end of Ehrlich's career in the clinical arena, which was mainly a result of the sudden death of Frerichs, should be more seriously considered a decisive break in Ehrlich's life than it has been thus far. He tried to use the laboratory, his most important working place, as a starting point to rebuild the old system. Under Koch, he had organized and performed clinical trials since 1895. But Ehrlich had to reorient himself; he was forced to pursue a secure position within the rigid framework of institutionalized German medicine, which was mainly university based. This was complicated by his Jewish faith, and he had to grasp Behring's and then Althoff's offers. Ehrlich remained indebted to the ideas of Julius Cohnheim on functional (experimental) pathology being related to clinical problems.[148] In Germany, where the older medical disciplines had been well established from about 1900, this left him between the chairs of pathology on one side and clinical disciplines on the other. Although loaded with problems, Ehrlich made strong efforts to initiate the testing of dyes with the help of clinical colleagues even after he had started to work intensely on the side-chain and receptor theory. These efforts were not successful, and in such a situation the theoretical evidence of his approach became more important than ever.

In his institute, Ehrlich concentrated on constructing a comprehensive theory, not only of immunological processes but also of human metabolism in general. The final aim continued to be the practical application on man, based on laboratory research, and his assistant Morgenroth as a keen follower of Ehrlich also performed patient-oriented immunological research work in his later career.[149] At first, Ehrlich was restricted to satisfying himself, the scientific community and the public with some vague assumptions about the future effectiveness of his concept in modern medicine. The critics of the side-chain and receptor theories, who raised their voices shortly after Ehrlich had developed his ideas, stimulated further consolidation of his theory, pulling Ehrlich more deeply into his pluralistic receptor world. Therefore, the development of the receptor concept depended on a combination of different events in Ehrlich's private and public life and in his academic career. These results correspond with recent findings of biographical research in general historiography, which have tried to explain the life of scientists based on their social and cultural environment.[150]

This new narrative helps us to understand why Ehrlich's receptor theory was not immediately generally accepted in pharmacology, and it complements the cited literature on the history of pharmacology.[151] Ehrlich's receptor concept was basic research, hard to understand, and the technical possibilities of his times were restricted. But there are further reasons: despite his reputation, Ehrlich was no pharmacologist and he deviated from the mainstream of contemporary pharmacology research. As we will see, the latter was devoted mainly to physical or physico-chemical views on the character of drug binding, based on the idea of a mechanical connection between dye-stuff and cell (see Chapter 3 below).[152] Above all, the doubts of contemporary pharmacologists were fuelled by the hypothetical character of Ehrlich's theory,[153] which could be pushed forward only with a vast propaganda apparatus. Although the side-chain theory provided the basis for Ehrlich's own research on cancer and chemotherapy[154] and was seen by colleagues as an inspiration for further work, there was at first no evidence of its usefulness for immunology or pharmacology in general.[155] As a construction which came into existence as a product of Ehrlich's social biography, personality and scientific career development, the receptors were the final stage of a process spiralling up into the enterprise of theoretical research. With his propaganda management and experimental system, Ehrlich was able successfully to adapt his hypothetical receptor system to every new challenge. The decision whether to join Ehrlich or to oppose him, whether to be a 'pluralist' or a 'unitarian', was similar to a religious confession. One had to believe in 'the theory' or to abandon it altogether.[156] It was very much shaped by its creator who tried to increase its credibility through a combination of persuasion and force.

But the receptor idea was not only promoted by Ehrlich and his collaborators but also by John Newport Langley, who is the subject of the next chapter.

2
The Development of the Concept of Drug Receptors in the Physiological Research of John Newport Langley

Whilst Paul Ehrlich was effectively the first to develop a receptor concept in the context of immunology, it was the Cambridge physiologist, John Newport Langley (1852–1925) who first proposed a receptor theory for the action of drugs and transmitter substances in the body. In this chapter we discuss how Langley developed, over a period of 30 years and in diverse research contexts, his ideas on the mode of action of drugs and physiological substances on tissues and cells. In 1905 these ideas culminated in the first full formulation of his concept of 'receptive substances' in cells.[1] We will also consider the influence of other British and Continental European scientists on Langley's thought and experimentation, and how his research themes were linked to other work in the Cambridge physiological laboratory and to the development of his academic career.

By the late nineteenth century animal experimentation had developed into the key method of physiology. Following anti-vivisectionist protests, experiments on living vertebrate animals were first regulated in Britain with the Cruelty to Animals Act of 1876.[2] While the use of anaesthetics in animal experiments was made a legal requirement (unless insensibility defeated the object of the investigation) and animal experimentation was restricted to licensed and inspected laboratories, this regulatory framework also provided a certain protection for compliant researchers. Langley experimented extensively on anaesthetized animals. This chapter illustrates how his ideas on drug receptors emerged from several different issues that were debated in experimental physiology at that time, and how they depended on specific physiological research techniques.

Langley's path to the receptor idea, as this chapter will make clear, was independent from that of Ehrlich. It will also be shown how Langley defended his concept against his scientific critics, and how he refined and elaborated upon the receptor idea during this process. Finally we will highlight how Langley not only used references to Ehrlich's side-chain theory to consolidate his own concept of receptive substances, but simultaneously asserted his intellectual independence from the German Nobel Prize winner.

Langley's early research under Michael Foster: drug effects on the heart and the salivary glands

J. N. Langley's path to a receptor theory originated in the research environment of the Cambridge School of Physiology under Sir Michael Foster (1836–1907). In October 1871 the 18-year-old Langley was admitted to St John's College, Cambridge, initially to study mathematics and history in preparation for a planned career in the Indian civil service. However, inspired by Foster's teaching, he changed direction in his second year of study and began to read for the natural sciences tripos. The young Langley was especially attracted by Foster's classes in elementary biology, embryology and physiology. Foster, who had been appointed as praelector of physiology at Trinity College in 1870, held to the principle of providing his students from early on with the opportunity to acquire first-hand knowledge through their own experimental work in his physiological laboratory, in those days a single room that was also used for lectures. Even before Langley graduated with a BA in 1875, Foster had involved him in his research programme. Langley was one of a small group of student-researchers who went on to scientific careers; among them were the physiologist Walter Holbrook Gaskell (1847–1914), the embryologist Francis Maitland Balfour (1851–82) and the physiological chemist Arthur Sheridan Lea (1853–1915).[3] When in 1876 Henry Newell Martin (1848–96), Foster's demonstrator, left for a professorship in biology at Johns Hopkins University in Baltimore, Langley was given his post. Langley assisted Foster in preparing a course in 'practical physiology', which included histology as well as physiological chemistry and animal dissection. Since the late 1850s Foster had been interested in whether the beat of the heart originated from the heart's nerve supply or from its muscular tissue, a question that was then widely debated among physiologists. In the 1870s, he and his students pursued this research problem more systematically. His animal experiments, performed on the hearts of snails and frogs, suggested that the origin of the heartbeat lay mainly in the muscular cardiac tissue itself.[4]

When Foster received a sample of the South American drug jaborandi from the London physician and physiologist Sydney Ringer (1835–1910), he passed it on to his student Langley for testing, particularly regarding its effects on the heart. In animal experiments Langley observed a significant slowing of the heartbeat after an extract of the drug had been injected. Such an effect could be interpreted as the consequence of a stimulating action of jaborandi on 'inhibitory fibres' of the vagus nerve ending in the heart. However, this slowing effect could still be produced after the Indian arrow-poison curare – which was widely held to paralyse nerve endings – had been administered. Langley therefore concluded that jaborandi acted 'probably ... more peripherally than the endings of the vagus nerves'.[5]

This view conflicted with that of the Paris physician and experimental pathologist Edmé Félix Alfred Vulpian (1826–81), who had proposed that jaborandi stimulated the endings of inhibitory vagus fibres in the heart, based on experiments in which curare appeared to prevent jaborandi's slowing effect on the heartbeat. Langley was therefore drawn into a more detailed examination of the drug's cardiac action. Experimentation on an anaesthetized rabbit confirmed his view that jaborandi acted on some structure other than inhibitory nerve fibres. In frogs he additionally investigated the antagonistic action of jaborandi and the alkaloid atropine on the heart. Applying solutions of the two substances directly to the heart of a frog whose brain and spinal cord had been destroyed, Langley showed 'that a definite quantity of atropia can only prevent a proportionate definite quantity of jaborandi from producing its effects on the heart' and 'that the condition of the heart ... depends on the relative amounts of jaborandi and atropia present'. Moreover, the antagonism between the two poisons could be demonstrated locally, by way of direct application, in different parts of the heart. These observations suggested to Langley that the drugs acted *directly* on the whole tissue of the heart – not on some localized nervous mechanism that caused and controlled the heartbeat.[6]

In this way Langley's work on jaborandi provided supporting evidence for Foster's view that the heartbeat had a muscular, not a nervous origin. Yet, this was not its only significance. At the very start of his career, Langley had hit upon a problem that would recur again and again in the course of his experimental work and that ultimately led him to his theory of receptive substances: do drugs act directly on the effector cells (in this case, the heart cells) or do they primarily affect the endings of nerves terminating in the organ tissues? Moreover, how do drugs combine with the tissues that they affect? And how do they cause changes in the cell's function? At the time, Langley believed that an action of jaborandi on muscle tissue alone, and not at all on nerve cells, was rather unlikely. But he recognized that the problem required further investigation.

In his next series of experiments Langley was able to use the alkaloid pilocarpine, which had been isolated from jaborandi bark and leaves in 1875. Jaborandi was known to produce salivation in human beings and higher animals. In experiments on the sub-maxillary salivary gland of the dog, Langley showed that pilocarpine and atropine acted as mutual antagonists with regard to salivary secretion. The secretion caused by pilocarpine could be stopped through atropine, restarted by pilocarpine, stopped again by atropine, and so on. Comparing this with his earlier findings he concluded: 'the secretion or absence of secretion is dependent on the relative quantity of the two poisons present, just as is the stand-still or beat of the heart'.[7]

A year later, in 1877, the Zurich physiologist Balthasar Luchsinger (1849–86) published an experimental study on the antagonism between pilocarpine and atropine on the secretion of sweat glands in the cat. Luchsinger gave a very graphic description of the mutual antagonism between the two alkaloids, stating that their actions summed themselves algebraically 'like wave crests and hollows, like plus and minus'.[8] Langley's interest was immediately aroused, as this appeared to be a parallel case to his own observations on the antagonistic effects of pilocarpine and atropine on the salivary secretion of the dog. During a stay at the laboratory of the Heidelberg physiologist Wilhelm Kühne, in 1877, Langley explored the matter further, performing experiments on the sub-maxillary gland of the cat. Here he found that the antagonism between the two drugs was not quite as simple as Luchsinger had described it. It was dose-dependent and thus incomplete. If large doses of atropine had been applied, pilocarpine could less fully produce secretion; and when very large doses of pilorcarpine were administered, this did not produce secretion, and this condition could not be antagonized by atropine.

Nevertheless Langley followed Luchsinger on an essential point. The Swiss researcher had concluded that the effect of the antagonism between the two alkaloids depended 'simply and solely upon the relative number of the poison molecules present' and that the antagonistic alkaloids bound chemically to the 'living protein' (protoplasm) of the cell. In this way, compounds between either the stimulating pilocarpine or the inhibiting atropine and the cell's protein molecules were formed, depending on the mass of each poison present and their relative affinity to the protoplasm.[9] While Langley was still unsure about the question of whether the poisons acted on the nerve endings in the salivary gland or on the gland cells themselves, he elaborated on Luchsinger's idea of a chemical union between drug molecules and cell components:

> ... we may, I think, without much rashness, assume that there is some substance or substances in the nerve endings or gland cells with which atropin and pilocarpin are capable of forming compounds. On this assumption then the atropin or pilocarpin compounds are formed according to some law of which their relative mass and chemical affinity for the substances are factors. In the analogous case with inorganic substances, other things being equal, these are the sole factors. To take the simplest case, if a and b are both able to form, with y, the compounds ay, by, then ay and by are both formed, quantity of ay and by depending on the relative masses of a and b present and their relative chemical affinity to y.[10]

Langley realized that in view of the incomplete antagonism between pilocarpine and atropine, the laws for the formation of inorganic compounds might be applicable to this case only with some modifications, but he was convinced that the law of mass action had been illustrated by his study.

In 1878 Langley received a Cambridge MA, having been awarded a fellowship at Trinity College in the previous year. His new interest in the theory of drug antagonism was soon fuelled by the publication of a paper by Würzburg pharmacologist Michael Joseph Rossbach (1842–94), who attacked the idea of a direct mutual antagonism between poisons. According to Rossbach, a tissue once paralysed by an alkaloid could not be restored to its former condition by applying another alkaloid. In experiments similarly carried out on the sub-maxillary salivary gland (of the dog), Rossbach had found that the stoppage of secretion through atropine could be overcome by physostigmine (the alkaloid from Calibar beans) only if the atropine dose had been small. He explained this with the hypothesis that the alkaloids had two points of attack on the gland: the nervous part and the glandular part. Small doses of atropine were thought to paralyse merely the nerve for secretion, the so-called chorda tympani, and to leave the gland cells unaffected. In this case, physostigmine could then still produce a flow of saliva by stimulating the gland cells. If, however, a large dose of atropine was given, nerve fibres as well as gland cells would be paralysed, so that physostigmine would then be unable to restore secretion.[11]

Langley, using pilocarpine instead of physostigmine as the stimulating agent, provided experimental evidence against Rossbach's view.

Experimenting on the sub-maxillary gland of an anaesthetized cat, he paralysed the chorda tympani by intravenous injection of atropine: electrical stimulation of the nerve then no longer led to salivary secretion. He then restored the secretion by injection of pilocarpine into the duct of the gland, and stopped the salivary flow again by giving another dose of atropine intravenously. According to Rossbach's interpretation, this second dose of atropine would have paralysed the gland cells. Nevertheless, Langley could again produce secretion by injecting more pilocarpine, and even after this flow had been stopped again by atropine, yet more pilocarpine could still restore it – while the chorda tympani remained paralysed throughout. On the basis of these findings Langley maintained his view that there was a mutual antagonism (within a certain dose range) between the two poisons, pilocarpine and atropine. They acted on the same tissue, forming 'chemical compounds' with it, and the result depended 'on their relative chemical affinity to the tissue and the mass of each present'.[12]

However, Langley neither embarked on a larger study of the binding of drugs to cells nor did he continue work on the origin of the heartbeat along the lines of his teacher Foster. Instead he made the physiology of glandular secretion (especially of the salivary glands) his first major field of research. He was active in this area until about 1890, combining morphological, physiological and chemical methods of investigation.[13] Langley, in this way, demarcated his personal territory of expertise, while simultaneously making full use of the spectrum of research methods then employed in the Cambridge Physiological Laboratory. In fact, his work on this theme was linked to his professional establishment as a university teacher and as a researcher.

In 1883 Langley was elected a Fellow of the Royal Society of London, and in 1884 he took up his new Cambridge posts as lecturer in natural sciences at Trinity College and as university lecturer in histology. Over the following years Langley gradually assumed the role of Foster's deputy in the physiological laboratory, as the latter became increasingly involved in the organization and political representation of science and higher education.[14] In addition to his work on glands, Langley also contributed to anatomical and histological studies in the then debated problem of localization of brain functions and in the degeneration of nerve tracts. In this field he collaborated with, among others, the later Nobel laureate Charles Scott Sherrington (1857–1952) and with Albert Sidney Grünbaum (1869–1921), who became director of the clinical laboratory of Addenbrooke's Hospital, Cambridge.[15] Thus equipped with a background in gland physiology as well as neuro-anatomy, Langley was prepared to enter a new research field which had been opened up by his colleague in the laboratory, Walter Holbrook Gaskell, lecturer in physiology since 1883.

Drug action and the autonomic nervous system

From the late 1880s Langley became interested in the vegetative or 'involuntary' nervous system, which had been examined both morphologically and physiologically by Gaskell. In the early 1880s Gaskell had demonstrated the existence of inhibitory as well as accelerator fibres in the vagus nerve of cold-blooded animals. This observation provided a unifying explanation for the wide range of effects on the heart that occurred with stimulation of the vagus nerve and that had puzzled earlier physiologists. Gaskell's subsequent research explored the hypothesis that it was not only the heart but all involuntary muscles that were innervated by two different, antagonistic types of visceral nerve fibres. By 1886 he had distinguished morphologically the visceral (vegetative) fibres stemming from the thoracic part of the spinal cord (that is, the sympathetic system) from the fibres that originated from its cervico-cranial and sacral regions (that is, the parasympathetic system). Moreover, Gaskell had pointed out that the actions of the thoracic part of the vegetative nervous system were antagonistic to the actions of the other two parts.[16]

Against this background, a methodologically important observation was made in 1889 by Langley and a medical collaborator, William Lee Dickinson (1863–1904) of Caius College, Cambridge. They found that nicotine selectively blocked nervous conduction in sympathetic ganglia, that is, that it interrupted the transmission of nerve impulses from the pre-ganglionic to the post-ganglionic nerve fibre. By electrically stimulating the nerve fibres running to and from a ganglion, before and after local application of a nicotine solution to the ganglion, it was possible to distinguish which fibres ended in the nerve cells of the ganglion and which simply passed through it.[17]

In the following years Langley used nicotine and other drugs as tools for a detailed functional and structural analysis of the sympathetic and parasympathetic systems, to which he gave the now common collective name 'autonomic nervous system'. His work in this area clearly promoted his professional career. In 1893 he became president of the Neurological Society of Great Britain and in 1899 president of the physiological section of the British Association for the Advancement of Science. As the new century started, Langley was an internationally recognized expert in the research field of the vegetative nervous system.[18] In 1896 he had been awarded an ScD, and when Foster retired in 1903, Langley succeeded him in the chair of physiology, having been deputy professor of physiology since 1900.[19] Back in 1894, he had already taken over the editorial responsibilities for Foster's *Journal of Physiology*. Langley's research on the sympathetic and parasympathetic nervous systems brought him to the top of the scientific establishment. In 1897–98 he served on the council of the Royal Society and in 1904–05 as its vice-president. After Foster's retirement from his professorship, Langley successfully managed the expansion of the facilities for physiology at Cambridge, culminating in 1914 in the opening of the new school of physiology, which had been built under his supervision with financial support from the Drapers' Company of London.[20]

The pharmacological issues that Langley had addressed in his early work on the heartbeat and salivary secretion emerged again in the 1890s in the new context of his research on the autonomic nervous system. A question linked with the paralysing action of nicotine on sympathetic ganglia was whether other poisons similarly affected nerve *cells*, or rather the *endings* of nerve fibres. From experiments on frog hearts, Langley and Dickinson concluded that nicotine acted upon nerve cells in the heart, whereas muscarine (the poisonous alkaloid of the fly agaric mushroom) and its antagonist atropine appeared to exert their effects on the peripheral endings of the vagus fibres leading to the heart.[21] In a further series of trials they tested several other alkaloids and poisons, including picrotoxin, apomorphine, codeine, cocaine, curarine, brucine and strychnine, in relation to their point of attack, that is, nerve cell body or nerve ending, by examining their effect on the sympathetic ganglia and nerve fibres in anaesthetized rabbits. More generally, Langley and Dickinson also hoped to uncover differences in the poisons' modes of action that might open up 'a new line of physiological investigation'. However, there were inconsistencies between the effects observed after local application to the nervous structures and after intravenous injection. Nicotine remained the clearest example of a poison that seemed to affect nerve *cells* (that is, the cell body) rather than the endings of nerve fibres.[22]

As the histological work of the Spanish anatomist Santiago Ramòn y Cajal (1852–1934), the Würzburg professor of anatomy, Rudolf Albert von Koelliker (1817–1905), and of Wilhelm von Waldeyer-Hartz provided evidence for a discontinuity between nerve endings and nerve cells (that is, for the neurone theory), Langley began to doubt his former interpretation and tended to believe that nicotine did not actually affect the nerve cells in the ganglion but the endings of the pre-ganglionic fibres that terminated close to them.[23] However, after cutting off the pre-ganglionic nerve fibres and allowing them to degenerate for up to 26 days, local application of nicotine to sympathetic ganglia still caused its characteristic effects. The result of this animal experimentation, reported by Langley in 1901, seemed to confirm the exceptional role of nicotine as a substance directly attacking nerve cells rather than the endings of nerve fibres. Though favouring the new neurone theory over the old concept of a continuous network of nerve fibres and nerve cells, Langley kept an open mind on this more general issue. He adopted the terminology of the new theory, 'because the facts cannot be expressed in terms of both theories without extraordinary verbiage'. Believing in the independence of the histological and physiological evidence, he held that even in the event that the neurone theory would have to be abandoned, his physiological observations would still remain valid.[24]

However, the issue of the precise point of attack of a drug or poison was brought up again at this time through new physiological research with extracts of the suprarenal gland (containing adrenalin). This pioneering work was undertaken by the Harrogate physician, George Oliver (1841–1915), and the professor of physiology at University College London, Edward Albert Schäfer (1850–1935). One observation made by Oliver and Schäfer, which became particularly important for Langley, was that suprarenal extract seemed to produce the typical rise in arterial blood pressure by directly acting on the smooth (unstriated) muscle tissue of the blood vessels. When Oliver and Schäfer added suprarenal extract to the perfused arterial system of frogs whose central nervous system had been destroyed, the small arteries (arterioles) contracted so much that the flow of the circulating perfusion fluid came almost to a standstill. Moreover, when the nervous plexus of the foreleg of an anaesthetized dog was cut on one side, but left intact on the other side, intravenous injection of suprarenal extract caused a prolonged diminution of size (measured with a plethysmograph) equally in both forelegs. This spoke for a direct action of the extract on the muscle tissue of the arterioles, rather than for an effect on nerves.[25]

German physiologists soon reported other experimental findings that supported the view that suprarenal extract acted directly on muscle tissue, not on nerve endings. Max Lewandowsky (1876–1918) of the Physiological Institute in Berlin, for example, showed in an experiment on the cat that the extract continued to produce contraction of the smooth muscles of the eye and eye-socket (innervated by sympathetic nerves) even after the sympathetic ganglia of the neck had been excised and the post-ganglionic nerve fibres had degenerated.

Moreover, by expanding upon experiments made by Oliver and Schäfer, the Göttingen physiologist, Heinrich Boruttau (1869–1923), demonstrated that suprarenal extract also acted directly on somatic striated muscle tissue – not on motor nerve fibres.[26]

For Langley, this all meant that there was apparently a second substance, besides nicotine, that *directly* affected cells rather than nerve endings. He repeated Lewandowsky's experiment on the cat's eye and was able to confirm his findings. He also returned to his tried and tested experimental model, the sub-maxillary gland of the cat, and found that suprarenal extract caused secretion even after the upper sympathetic ganglion of the neck had been removed and ten days had been allowed for the post-ganglionic secretory nerve fibres to degenerate. These results spoke for a direct action on the effector cells, that is, the muscle cells and gland cells, respectively. Moreover, Langley was probably the first in this context to point to the striking parallels between the effects of suprarenal extract and of electrical stimulation of sympathetic nerves. This latter observation placed him in the early history of research into a chemical transmission of nerve impulses – a field that became very prominent in physiology in the first half of the twentieth century.[27]

However, although there were arguments in favour of the direct binding of drugs and physiological substances to muscle or gland cells, difficulties remained. Remarkably, these difficulties induced Langley to speculate further about the nature of drug binding to cells. First, there were considerable differences in the effect of suprarenal extract on different tissues innervated by autonomic nerve fibres. Second, differences occurred also in the effect of nicotine on sympathetic ganglia. It paralysed the upper ganglia of the neck and the ganglia of the lateral chain more easily than those of the solar plexus. Third, there were differences in nicotine's efficacy in related species, such as dogs and cats, and even between individual animals of the same species. Langley speculated that the reaction to nicotine depended 'upon the presence of a special chemical substance in the nerve-cells or on the nerve-endings', and that there had to be differences 'in the chemical constitution of protoplasm' that were responsible for the differences in the alkaloid's efficacy in different animals. Without naming Paul Ehrlich, or any other researcher in the nascent field of immunology, Langley drew attention to 'recent investigation[s] upon toxins and anti-toxins', which had shown 'what enormous effects on the organism these differences in chemical constitution may bring about'.[28]

Elliott's 'myoneural junction' and Langley's 'receptive substances'

In the meantime, Langley's student, Thomas Renton Elliott (1877–1961), had taken up his professor's observation of the parallel actions of suprarenal extract and sympathetic nerve stimulation. Working with 'adrenalin', which had been isolated from suprarenal glands by Jokichi Takamine (1854–1922) in 1901, Elliott provided further examples of 'the broad rule that the action of the substance [suprarenal extract] upon plain [that is, smooth] muscle simulates that of electrical excitation of the sympathetic nerves supplying each particular muscle'.[29] He also investigated some apparent exceptions to this rule. In May 1904, in a preliminary communication to the Physiological Society, Elliott made the suggestion that adrenalin was 'secreted by the sympathetic paraganglia' and might be 'the chemical stimulant liberated on each occasion when the [nervous] impulse arrives at the periphery'.[30] While this proposal was another crucial step in the development of the concept of chemical neurotransmission, his thoughts on how the muscle cell received the stimulus of the 'chemical excitant' and reacted with a change of tension of the muscle fibres were important for the development of the receptor idea.

Langley and Lewandowsky's view that suprarenal extract acted directly on muscle cells had been criticized by Thomas Gregor Brodie (1866–1916), professor-superintendent of the Brown Institution, and Walter Ernest Dixon (1870–1931), then assistant to the Downing Professor of Medicine in Cambridge. Reporting their own experiments as well as experimental results obtained by other researchers, Brodie and Dixon argued that adrenalin affected the sympathetic nerve endings, not the peripheral tissues themselves. In particular they suspected that Langley and Lewandowsky had not allowed enough time in their degeneration experiments for the sympathetic nerve endings to completely disappear before the suprarenal extract was tested. Elliott addressed this criticism. In an experiment on a cat he excised on one side the sympathetic ganglia connected with the eye. Nearly ten months later he tested the effect of an intravenous injection of adrenalin on the animal's iris: the typical dilation of the pupil appeared even more quickly and was more extensive on the operated side than on the normal side. In Elliott's mind there was no doubt that the sympathetic nerve endings had entirely degenerated after such a long time and that the adrenalin therefore could not have acted on them.[31]

Another critical argument proposed by Brodie and Dixon came from an experiment that Dixon had performed with apocodeine. He had shown that the contraction of the muscle tissue of blood vessels that was typically produced by adrenalin could be almost completely prevented by prior injection of apocodeine. Yet, subsequent injection of barium chloride still led to constriction of the vessels. This meant (for Brodie and Dixon) that the muscle tissue as such had not been injured by the apocodeine, and that therefore both apocodeine and adrenalin had acted on the nerve endings terminating in the blood vessels.[32]

Against this background of conflicting experimental evidence Elliott suggested that it was neither the nerve endings nor the contractile fibres of the muscle cell that were affected by adrenalin. Instead he proposed that the 'substance' that was excited by adrenalin was the 'myoneural junction', that is, the link between nerve ending and muscle cell, which he believed to originate from, and to be sustained by, the muscle cell. This hypothesis also explained why adrenalin affected only those tissues that had a sympathetic innervation. The union with sympathetic nerves during phylogenetic development, he believed, had led to the growth of a special 'substance' in the muscle cells that could be excited by adrenalin. The nature of this substance, that is, of the myoneural junction, determined whether the impulse travelling down a sympathetic nerve led to contraction or inhibition (relaxation) of the muscle fibres. In this way the differences in the action of adrenalin in different tissues could be explained. Moreover, Elliott speculated that the other parts of the autonomic nervous system, that is, the parasympathetic nerves and the autonomic ganglia, and also the skeletal nerves leading to striated muscle, had a different type of junction from that in the sympathetic nerves.[33] From these considerations it was only a very small step for Langley, who had probably guided his student's thoughts in this difficult and controversial matter,[34] to formulate his concept of 'receptive substances'.

In December 1905, in his paper on receptive substances, Langley critically reviewed the evidence that had been provided on the direct action of certain drugs and poisons on cells. In particular the experiments involving degeneration of the nerve fibres before the effects of a drug were tested, as carried out by Lewandowsky, Elliott and himself, appeared crucial. In light of Brodie and Dixon's criticism, Langley gave details of another experiment of his own on a cat, in which he showed that an extract made from 'Burroughs and Wellcome's supra-renal tabloids' produced typical adrenalin effects on the head (eyes, sub-maxillary gland and so on) even fourteen and a half months after the upper sympathetic ganglion of the neck had been excised and when the sympathetic nerve fibres coming from it should long since have degenerated. The point that certain poisons acted directly on cells, for example, nicotine on nerve cells, or adrenalin, pilocarpine and atropine on smooth muscle cells and gland cells, seemed now to him quite well established.

There remained, however, two problems that required a more differentiated account of the drugs' mode of action. The first was that the efficacy of adrenalin on smooth muscle differed considerably between various tissues in the body, even between tissues that were innervated by the sympathetic nervous system. The second problem was Dixon's finding that apocodeine prevented the vascular constricting action of adrenalin, but not that of barium chloride, which suggested that adrenalin acted on nerve fibres, not on muscle fibres. In response to these problems Langley proposed that adrenalin did not directly stimulate the muscle cell's 'contractile substance *quâ* contractile substance', but that it acted on 'accessory protoplasmic substances' of the cell. Intrinsic differences in these accessory substances could explain the differences in the efficacy of adrenalin in various (smooth) muscle tissues.[35]

Langley next turned to nicotine and its effect on striated, skeletal muscle to further support this hypothesis. In an anaesthetized fowl an intravenous injection of nicotine caused a characteristic prolonged, tonic contraction of the gastrocnemius muscle of the leg, even after the sciatic and internal peroneal nerve had been cut in order to exclude any central nervous influence. Also when the internal peroneal nerve had been 'paralysed' through nicotine (that is, when electrical stimulation of the nerve no longer led to a muscular contraction), a larger dose of nicotine still caused the gastrocnemius muscle to contract. This indicated that nicotine acted directly on muscle cells. Intravenous injection of curare abolished the nicotine-induced tonic contraction, and further injection of nicotine brought it on again, that is, the two poisons acted as mutual antagonists. As Langley pointed out, the relation between nicotine and curare was the same as the relation between pilocarpine and atropine, which he had described 27 years earlier in his experiments on the submaxillary salivary gland. Accordingly, he suggested that nicotine and curare acted upon the same 'protoplasmic substance or substances' of the muscle cell. Whether these substances combined predominantly with nicotine (resulting in stimulation) or with curare (leading to relaxation) depended on 'the relative amount of the two poisons' present.[36]

In a subsequent series of experiments Langley cut the peroneal nerve, excised a piece of it, and allowed periods between 6 and 40 days for the peripheral part of the nerve to degenerate. Functional regeneration was excluded in tests with electro-stimulation of the proximal part of the cut nerve, and the degeneration of the nerve endings was confirmed in histological examinations. Yet, injection of nicotine still produced the typical tonic contraction with the responsiveness of the muscle to the poison actually being increased, and curare still exerted its antagonistic effect on the nicotine contraction. Moreover, direct electrical stimulation of the muscle after injection of nicotine or after injection of curare could still produce some contraction.[37]

From these observations Langley drew the critical conclusion that 'neither the poisons nor the nervous impulse acted directly on the contractile substance of the muscle but on some accessory substance', and he continued: 'Since this accessory substance is the recipient of stimuli which it transfers to the contractile material, we may speak of it as the *receptive substance* of the muscle.'[38]

Referring briefly to Ehrlich's side-chain theory of immunity, Langley speculated that a receptive substance might be 'a side-chain molecule of the molecule of contractile substance'. He remained cautious though, adding that to him there seemed to be no advantage at present in 'attempting to refer the phenomena to molecular arrangement'. However, having produced evidence for the action of adrenalin as well as of nicotine and curare on 'accessory' or 'receptive' substances of the cell, Langley dared to generalize. He suggested that alkaloids such as pilocarpine, atropine and strychnine also acted in this manner, as might other, internally secreted substances (that is, hormones), such as secretin, thyroidin and 'the various stimulating chemical bodies formed by the generative organs'.[39] Moreover, Langley proposed as a rule:

> So we may suppose that in all cells two constituents at least are to be distinguished, a chief substance, which is concerned with the chief function of the cell as contraction and secretion, and receptive substances which are acted upon by chemical bodies and in certain cases by nervous stimuli. The receptive substance affects or is capable of affecting the metabolism of the chief substance.[40]

With these conclusions Langley had laid the foundations for a theory of drug receptors in cells. Significantly though, Langley located his 'receptive substances' *in* the cell rather than *on* the cell. In this respect his receptor concept was different from the modern one, which describes receptors within the cell as well as in the cell membrane.

Langley's 'receptive substances' had similarities with Ehrlich's 'side-chains' that would bind bacterial toxins to the cell. But when Langley formulated his concept of receptive substances in 1905, Ehrlich still believed that the side-chain theory was applicable only to toxins and bodily substances, not to drugs, chiefly because drugs did not seem to be firmly fixed in the tissues and could easily be washed out of them by solvents (see Chapter 1). Langley, on the other hand, assumed that a chemical union between a cell's receptive substances and a drug was formed. Returning to the analogy of binding in inorganic chemistry that he had used in his discussion of pilocarpine and atropine, he spoke of the formation of 'nicotine-muscle compounds' and 'curare-muscle compounds'. Which of these two kinds of compounds prevailed depended 'upon the mass of each poison present and the relative chemical affinities for the muscle radicle [that is, the receptive substance or side-chain]'. Moreover, he speculated that the biological effect of either contraction (through binding of nicotine) or inhibition (through binding of curare) was caused by different 'chemical re-arrangements set up in the muscle molecule by the combination of one of its radicles'.[41]

It was only in 1907 and especially in response to Langley's work on receptive substances and alkaloids that Ehrlich changed his mind and proposed the existence of 'chemoreceptors' for drugs.[42] Langley's receptor concept also had considerable similarities with Elliott's concept of the 'myoneural junction' on which a nervous impulse or chemical stimulant would act. In fact, Langley acknowledged that Elliott's work on adrenalin had 'made the issues clearer' for him. He also agreed with Elliott's hypothesis that it was the nature of the myoneural junction that determined whether a nerve impulse resulted in contraction or inhibition. Langley suggested that a cell could make two kinds of receptive substances, 'motor' and 'inhibitory', and that the effect of a nervous impulse on the cell depended on the proportion of these two receptor types.[43]

However, he disagreed with his student about how the receptive substances may have been formed and how they had obtained their characteristics during phylogenesis. According to Elliott, in developing a union with nerve endings, cells had grown a specific myoneural junction. On this supposition Langley expected that in the nerve-degeneration experiments the myoneural junction, or the receptive substance, degenerated as well, leading to a diminished physiological response to the application of drugs. But as the experiments with adrenalin and nicotine had shown, cells were even more sensitive to the drugs after denervation. Langley had also performed some experiments with adrenalin, nicotine, and strychnine on chicken embryos – obviously based on the common theory that ontogenesis represented a brief repetition of phylogenesis.

In these experiments the drugs showed quite marked effects in very early developmental stages. This was a finding that did not support Elliott's suggestion. Finally, the variety of effects caused by sympathetic nerve stimulation and adrenalin in various tissues, and the incomplete parallelism between the two, spoke, in Langley's view, against Elliott's explanation. As Langley saw it, the various body cells had a constant tendency to vary in their chemical composition, which upon the formation of a functional connection with a nerve at some point in phylogenesis had merely become 'fixed'. Different parts of the nervous system formed their connection with the peripheral tissues at different periods of phylogenetic development. In this way different types of receptive or 'synaptic' substances had been established.[44]

Yet Elliott did not agree with this interpretation. In a subsequent study on the nerve supply of the bladder in various animal species, he called Langley's theory of receptive substances 'a doctrine of inflexibility'. In particular, he claimed that Langley had attributed too little influence to the nature of the nerves that entered the tissues during phylogenesis and had put too much emphasis on independent chemical changes of the peripheral (muscle) cells. As Elliott put it, Langley's view did 'not clearly ascribe a determinant value to the entering nerve, which must knock patiently unheard until the cell chances to develope [sic] the proper receptive substance'.[45] However, by the time this criticism was published, in 1907, Elliott had already left Cambridge for his clinical education at University College Hospital, London, and this particular debate between the two researchers seems to have been discontinued. Elliott went on to become assistant physician to the hospital in 1910 and continued his research with work on the functions and nervous control of the suprarenal glands. After the First World War he became professor of clinical medicine at UCH.[46]

Criticisms and further development of Langley's receptor theory

On 24 May 1906 Langley gave the Croonian Lecture to the Royal Society on the topic of his new concept of receptive substances, adding some more experiments with nicotine and curare, made on the frog and toad.[47] Immediately afterwards he visited Europe to further disseminate his ideas. On 28 May he spoke to the Morphological-Physiological Society of Vienna about 'nerve endings and special receptive substances in cells'.[48] In the following year, Langley presented his receptor concept at the Seventh International Congress of Physiologists in Heidelberg.

Reporting experiments with local application of nicotine solutions to various muscles of the frog, he elaborated on his evidence for different kinds of receptive substances. The results of these trials, which also included tests on denervated muscles and with the antagonist curare, indicated that the frog muscle had at least two types of receptive substances for nicotine: one leading to a slow and prolonged ('tonic') contraction, and the other causing a rapid and brief contraction ('fibrillar twitching'). Both types could be located in the region of nerve endings as well as in other parts of the muscle fibre. Since local application of the alkaloid veratrine caused yet another pattern of contraction, Langley presumed that there had to be further types of receptive substances in the muscle. In general, he considered these substances to be radicles of the contractile molecule of the muscle cell, and he suggested that those near the nerve endings might have undergone a special development.[49]

However, Langley's arguments for the existence of receptive substances in cells quickly encountered the criticism of other researchers. At the same congress of physiologists he was confronted with a critical paper by one of his former collaborators, Rudolf Magnus (1873–1927), who was then a lecturer in the pharmacological institute of Heidelberg University.[50] Magnus focused on one key argument of Langley's for receptive substances: the mutual antagonism of nicotine and curare on the denervated muscle. Magnus acknowledged the general validity of the method of testing poisons after degeneration of nerves in order to establish whether or not they acted on the peripheral tissues or on nerve endings. But he was not convinced that the mutual antagonism between two poisons actually indicated their specific point of attack. Langley had concluded that curare, like nicotine, bound to receptive substances of the muscle cells, because he had found that curare abolished the nicotine-induced contraction of the denervated muscle. Magnus argued against this observation with findings from his own experiments on various muscles of the rabbit, in which the relevant nerves had been cut and allowed to degenerate. In these trials he used physostigmine instead of nicotine as the stimulant agent and antagonist of curare. He found that physostigmine failed to produce a contraction in the denervated muscle from the twenty-seventh day after section of the nerves. This spoke for an action on nerve endings, and according to Langley's logic, the antagonist curare would therefore also act on nerve endings. This example showed, according to Magnus, that the conclusion about the point of attack of curare depended on which antagonist had been used. If one used nicotine, as Langley had done, the evidence suggested that curare acted on the muscle cell. If one used physostigmine, like Magnus, one was led to conclude that curare affected the nerve endings. In other words, 'nothing at all' could be found out about a poison's point of attack from trials with its antagonists.[51]

Magnus admired Langley's work in physiology, especially his outstanding skill in animal experimentation. Unsurprisingly, therefore, Magnus politely emphasized that he did not wish to criticize Langley's doctrine of receptive substances as such. However, he made it clear that one of the Cambridge professor's 'proofs' for it, the mutual antagonism between nicotine and curare on the denervated muscle, was questionable.[52]

In the discussions following Langley's and Magnus's presentations, Langley suggested that the receptive substances of the denervated rabbit muscle had degenerated in addition to the nerve endings, which explained why Magnus had no longer obtained a contraction on injection of physostigmine. Yet this argument constituted a certain contradiction of Langley's earlier observation that denervated muscle cells actually showed an increased sensitivity to drugs such as nicotine and adrenalin. On the other hand, Magnus had to admit that curare might have two points of attack: the nerve ending and the muscle cell.[53] By the following year both researchers had collected more evidence to support their divergent points of view. Magnus argued from experiments conducted by himself, by Langley's Cambridge collaborator Hugh Kerr Anderson (1865–1928) and others that similar inconsistencies resulted in the point of attack of atropine if one drew conclusions from its antagonistic action to pilocarpine and physostigmine.[54] Langley demonstrated different kinds of contraction after nicotine and after physostigmine had been applied to muscle, and concluded from this that there had to be 'different receptive radicles' for the two poisons.[55] In this way, Magnus's criticism led eventually to a greater complexity of Langley's receptor concept.

Langley also had to consider the recent results of Hermann Fühner (1871–1944) of the pharmacological institute in Würzburg, who had studied the muscular effects of the curare-like substance guanidine for his habilitation thesis. Fühner had found that guanidine chloride failed to produce the usual contraction ('fibrillar twitching') of the frog's gastrocnemius when it was applied to the muscle 11 and 13 days after its nerve had been cut. For the Würzburg researcher this suggested an action of guanidine (and, by extension, of curare) on nerve endings, which would have degenerated by that time. Yet, he obtained some guanidine contractions again from the sixteenth and eighteenth day onwards and explained this by proposing regeneration of the nerve endings. In Langley's view, Fühner's hypotheses about degeneration and then regeneration in the absence of a connection to the central nervous system were untenable. Moreover, Langley referred to denervation experiments on another leg muscle, the sartorius, which showed with histological staining that the nerve endings needed about six weeks to degenerate and did not begin to regenerate until the sixty-ninth day. Accordingly he did not accept Fühner's evidence for guanidine's action on nerve endings rather than on muscle cells. Still, Fühner's experiments illustrated the uncertainties inherent in the nerve-degeneration method, and in this way cast doubt over another element in Langley's argumentation for the existence of receptive substances in cells.[56]

Criticism came not only from German researchers but also from colleagues at home. Stimulated by Langley's ideas on receptive substances and Ehrlich's on chemoreceptors, Walter Ernest Dixon, who had become lecturer in pharmacology at Cambridge, published a study of the specific action of strychnine on the spinal cord in 1909. Working with emulsions of spinal cord, he and Philip Hamill (1883–1959), his collaborator in the pharmacological laboratory, did not find any evidence for a chemical combination of the alkaloid with the nervous tissue. They therefore questioned the existence of specific receptors for vegetable alkaloids.[57] Doubts about Langley's concept of receptive substances were also raised from a chemical point of view. George Barger (1878–1939) and Henry Hallett Dale, then working at the Wellcome Physiological Research Laboratories, showed that a wide range of structurally differing amines apparently mimicked the physiological effects of sympathetic nerve stimulation. Moreover, they could not identify a common structural component that was specific for these 'sympathomimetic' amines. On these grounds they were sceptical about Langley's suggestion that drugs entered into chemical combinations with *specific* receptive side-chains of the cell.[58]

Perhaps the more enduring challenge to Langley's concept of receptive substances arose, however, from a new theory on the mode of action of drugs, which was developed by the Freiburg pharmacologist, Walther Straub. Inspired by studies of drug absorption undertaken by his academic teacher Rudolf Boehm in Leipzig,[59] the young Straub extended this line of research during a stay at the Zoological Research Institute in Naples in the spring of 1905. His experimental model was the isolated heart of the sea snail, *Aplysia*, on which he examined the effects of muscarine and its antagonist, atropine. Muscarine typically caused a slowing of the heartbeat. Straub concluded from his experiments that this effect occurred only as long as the poison entered the heart cells. The effect did not depend, in his view, on an action of muscarine within the cell itself. Rather, the important factor was the gradient in the poison's concentration between the outside and the inside of the cell membrane, or the 'concentration potential' as he called it, which kept the process of absorption in motion. After the cells had become saturated with muscarine, a further increase of its concentration outside the cell had no further effect. On this basis Straub developed a 'poison-potential theory' (*Potentialgifttheorie*): while the poison was entering the cell, the membrane was unable to excrete the cell's waste products. These accumulated and damaged the cell, leading to cessation of its functions. In Straub's opinion, this theory was not only applicable to muscarine but to other alkaloids, such as pilocarpine, physostigmine and nicotine, as well as to the hormone adrenalin. Moreover, the antagonism between muscarine and atropine could be explained with the hypothesis that atropine slowed the absorption of muscarine into the heart cells.[60]

This essentially *physical* theory of drug action stood in marked contrast to the concept of specific chemical binding of drugs to receptive sidechains as proposed by Langley and (subsequently) Ehrlich. As Straub put it rather bluntly in his Freiburg inaugural lecture in 1908, any remarks on a direct relationship between chemical structure and physiological effect of a drug were mere speculation.[61] Langley took Straub's observations seriously, but provided an explanation for them that was in harmony with his chemical theory. As long as the poison combined chemically with the receptive substances they 'set up a stimulus' to the cell. When they were saturated, there was no more stimulus and thus no further effect. Similarly, the antagonism between atropine and muscarine could be explained with the hypothesis that atropine combined with the receptive substances and in this way prevented the effect of muscarine.[62]

Langley also used the similarities with Ehrlich's side-chain theory to support his own concept of receptive substances. He interpreted these as 'atom-groups of the protoplasm' of the cell. Two such atom-groups had to be distinguished: the 'receptive' and the 'fundamental'. When chemical substances bound to the receptive atom-groups, they would alter the protoplasmic molecule of the cell and in this way change the cell's function. In less differentiated cells these atom-groups could also split off from the cell and act as antibodies in the blood, as suggested in Ehrlich's theory of immunity (see Chapter 1 above). In more differentiated cells, such as those of the muscles and glands, the receptive atom-groups had undergone a 'special development' which enabled them to combine with hormones or with alkaloids. Due to those cells' connection with nerve fibres, these further developed atom-groups tended to concentrate in the region of the nerve endings. The 'fundamental' atom-groups, by contrast, were essential for the cell's life. If a chemical substance bound to such a group, the cell would be damaged and die. Langley pointed out that this latter type of atom-group had been demonstrated by Ehrlich in recent experiments with arsenic compounds on trypanosomes (that is, the protozoa causing sleeping sickness). If arsenic bound to the chemoreceptors of the trypanosomes, these microorganisms were destroyed.[63]

By making this distinction, Langley used Ehrlich's side-chain theory to bolster his own concept of receptive substances. In 1908 Ehrlich was awarded the Nobel Prize for his studies into immunity. As Langley put it, his hypothesis of receptive substances constituted 'an extension of Ehrlich's side-chain hypothesis'. Langley preserved the originality of his own research by making it clear that Ehrlich's recent studies into drug binding were concerned with a different type of receptor to the one that he had examined in his experiments with nicotine and curare. Langley emphasized that he had arrived at his own receptor concept 'by entirely different experiments and by a different line of argument' and that he had proposed the binding of drugs to side-chains at a time (1905–06) when Ehrlich had not yet considered this possibility. Significantly, Langley neither adopted the term 'receptor', which Ehrlich had introduced in 1900 in the context of his immunological research, nor Ehrlich's neologism 'chemoreceptors' of 1907. He continued to employ his own term 'receptive substances'.[64]

One of Langley's postgraduate students, Archibald Vivian Hill (1886–1977) (a future Nobel Prize winner) provided further evidence in 1909 for his professor's chemical receptor theory and against the 'physical view'. Hill took a different approach to the problem by performing a quantitative and mathematical analysis of the contractions produced by nicotine, and the relaxation caused by its antagonist curare, in the frog's *rectus abdominis* muscle. He also examined these physiological effects at different temperatures. The formulas at which he arrived led him to the firm conclusion that nicotine as well as curare formed reversible chemical combinations with a constituent of the muscle.[65] Langley's predominantly qualitative evidence for the existence of receptive substances in cells was thus endorsed by an analysis of quantitative data. Nevertheless, the physical theory of drug action as introduced by Straub became a strong competitor of the 'chemical' drug receptor theory, and remained so until the 1930s (see Chapters 3 and 5 below). Langley acknowledged further developments in a physical theory of the specific action of poisons (for example, theories about differences in the permeability and solvent power of the cell membrane), but stayed committed to his concept of receptive substances and to a chemical theory of drug effects.[66]

In 1921, four years before his death, Langley published a final synthesis of his research into the autonomic nervous system. His views on receptive substances in cells had not changed:

> The known physical characters of drugs are insufficient to account for the effects they produce, though they account for a difference in rate of action; in consequence I consider that there is a chemical combination between the drug and a constituent of the cell – the receptive substance. On the theory of chemical combination it seems necessarily to follow that there are two broad classes of receptive substances; those which give rise to contraction, and those which give rise to inhibition.[67]

Conclusions

This chapter has illustrated the complexities that were involved in Langley's conceptualization of the drug receptor. His path to a receptor concept of pharmacological action was neither straightforward nor the result of a specific research plan. Langley's ideas about the interaction of drugs and poisons with cells developed intermittently over a period of 30 years and in diverse research contexts. The main contexts were the physiology of the heart, of salivary glands and, most importantly, of the autonomic nervous system. In addition we can identify a number of subsidiary contexts, such as the emerging theory of mutual antagonism of drugs, Ehrlich's side-chain theory of immunity, early hormone research, the beginnings of the neurone theory, and the first ideas about a chemical transmission of nerve impulses.

Gerald Geison has argued that the Cambridge School of Physiology, and by extension English physiology in general, was characterized (in comparison to German and French physiology) by especially close links between histological and physiological work, by a Darwinian evolutionary perspective on physiological problems, and by a theory that favoured the muscular tissue of the heart over its nerves in the origin of the heartbeat.[68] Langley's research on receptive substances in cells, as discussed in this chapter, reflects these general characteristics. His receptor concept was built on a combination of physiological and histological research methods; he tried to explain the diversity of receptor types through evolutionary processes; and he identified muscle tissue rather than the nerve endings as the site of action of alkaloids and hormones.

While our historical reconstruction allows us to recognize and follow the logic of Langley's personal intellectual route to his concept of receptive substances, it also identifies the various influences from other British and Continental European scientists upon his thought and experimentation. In retrospect, Elliott's ideas about the action of adrenalin on the 'myoneural junction' appear to have been especially relevant in Langley's final steps towards his receptor concept. Some of the experimental methods employed by Langley in proving the existence of receptive substances were contested at the time. As we have seen, fundamental questions were raised about the validity of antagonistic drug trials, and there were considerable uncertainties involved in the method of nerve degeneration. The question of whether antagonistic drugs such as nicotine and curare, or pilocarpine and atropine, acted on nerve endings or directly on the effector cells (muscle cells and gland cells), remained unanswered for many researchers despite the experimental evidence provided by Langley.

These methodological problems provide one explanation for subsequent difficulties in the recognition of the receptor concept. Moreover, the fact that the concept of 'receptive substances' had been developed in physiology, rather than in pharmacology, may have constituted an obstacle. Clinicians were sceptical about the practical significance of experimental physiologists' research. As Sir Charles Sherrington recalled, when the Glasgow professor of the practice of medicine, William Tennant Gairdner (1824–1907), made a visit to Cambridge and found Langley studying the salivation of the cat and Gaskell investigating the tortoise heart, his comment was: 'devoted laboratory work, but as regards medicine sadly beside the mark'.[69] But perhaps the most important reason for such difficulties in recognition was the direct competition between Langley's (and later Ehrlich's) *chemical* ideas about drug binding and *physical* theories of drug action, such as Straub's. Despite Hill's early quantitative evidence in favour of Langley's receptive substances, the controversy over a chemical versus a physical effect of drugs on cells remained unresolved after Langley's death in 1925. The extensive debates on this topic will be analysed in the following chapter.

3
Receptors and Scientific Pharmacology I: Critics of the Receptor Idea and Alternative Theories of Drug Action, *c.* 1905–35

By 1905 the receptor concepts of both John Newport Langley and Paul Ehrlich were fully developed, and in 1907 the two researchers shook hands on the special case of drug-binding receptors. Despite early resistance to certain aspects of their theories, Langley and Ehrlich were successful in promoting and publicizing their concepts within the scientific communities of physiology, immunology and bacteriology, and the new theories of drug-binding receptors were at least being considered by representatives of pharmacology. In the following two chapters we will concentrate on this last medical field. These chapters will describe and analyse the response of pharmacologists to the receptor concept between 1905 and 1935, a period that was characterized by discussions on the direct effect of drugs on cells. These debates were fuelled to a large extent by the receptor concept. Most pharmacologists were critical and a number of alternative theories emerged and were discussed. Within this period of transition, there was a break around 1930, when the Edinburgh pharmacologist Alfred Joseph Clark (1885–1941) revived the interpretations developed by Langley and Ehrlich and presented the receptor concept on the basis of a new approach (see Chapter 5 below).

This chapter deals with alternative theories of drug action which competed with the receptor concept during the first three decades of the twentieth century. In the first section we will discuss the heritage of nineteenth-century pharmacology, which shaped the scientific approach of the discipline's representatives until at least the end of the Second World War. This is a prerequisite to understanding the development of the so-called 'physical theory of drug action' as the most important alternative theory to the receptor concept. This will be the focus of the following two sections: the first concerns leading researchers in pharmacology; the second explores the general acceptance of the physical theory in this field. We will then give two examples of outstanding theories of drug action during this period, before summarizing and analysing our findings.

The beginnings of scientific pharmacology and its legacy in the twentieth century

Until the middle of the nineteenth century, the study of drugs was characterized by an empirical search for adequate therapies. The theory of drug application depended on the relevant medical system, and practices ranged from careful use of remedies (or even avoidance of drugs at all) to polypharmacy and prescription of heroic doses. In 1850, countless drugs were used on an empirical basis, but their mode of action within the human body was largely unknown. After about 1850, the field hitherto called 'materia medica' was reformed and transformed into the analytical and experimental discipline of 'pharmacology' which, in the context of a new, 'scientific' medicine emerging from about 1850, can be called 'scientific pharmacology'. Its basis was the exact analysis and classification of all therapeutic substances; effective and ineffective compounds had to be distinguished in order to improve the efficacy of drug therapy. In a further step it was necessary to determine as precisely as possible the qualitative and quantitative effects of the effective remedies. In relation to therapeutic medicine this was the search for and analysis of 'pure' substances and the estimation of the exact 'dose' causing a specific effect. This was done with the help of animal experimentation.[1]

At the forefront of the scientific community of nineteenth-century German pharmacology stood Rudolf Buchheim and his student Oswald Schmiedeberg. Buchheim had developed the programme of modern *experimental* pharmacology at the University of Dorpat. In 1847 he was appointed extraordinary professor and in 1849 full professor of pharmacology (*Diätetik und Materia Medica*). Buchheim's years in Dorpat between 1847 and 1867, when he moved to the University of Gießen, coincided with the period of increased understanding of scientific medicine. The pioneering spirit of modern medicine also impacted on Buchheim, who supervised 86 MD theses investigating in detail the constituents and effects of drugs.[2] It was in particular the systematic usage of animal experiments that nineteenth-century pharmacologists such as Carl Binz (1832–1913) considered to be the decisive improvement within the field.[3]

In 1876 Buchheim explained that knowledge about therapeutic substances could no longer be obtained purely through the observation of patients. The mode of action of drugs could only be detected by experimental studies in the lab, which had to be carried out by the pharmacologist as the specialist in the field.[4] According to Buchheim, the effects of a drug could best be studied by isolating its effective components and investigating their chemical properties, and then by correlating the chemistry of the drug with any changes caused by it in the function of organs; these correlations would increase the efficacy and safety of the application of any therapeutic substance. Importantly, the correlation between drug chemistry and drug activity had to be examined using the same methods and tools that were employed for physiological investigations.[5]

Although chemical knowledge and close links with clinical colleagues were important, the physiological education of the pharmacologist was – according to Buchheim – the most relevant area of expertise for questions relating to the mode of action of drugs. The pharmacologist had to carry out animal experiments or self-experiments in order to measure changes in the function of individual organs and/or the reactions of the whole organism, after the application of certain substances, using instruments or conducting chemical analyses of body fluids. The missing link in Buchheim's experimental chain was direct evidence of chemical changes within the *cells* of the organism, or evidence of the precise fate of the applied substance in the body. Although he systematized current knowledge about drugs, his experimental approach merely compared input and output.[6] He promoted this approach whilst keeping an eye on pharmacotherapy, which he hoped to improve.

Buchheim's approach did not cause a therapeutic revolution, but it equipped pharmacology with some sound methodology. His pupil Schmiedeberg helped to spread Buchheim's method of drug research, and in this way Buchheim became the leading authority in nineteenth-century pharmacology – at least in Germany. Like his teacher, Schmiedeberg based his pharmacological studies on chemistry, even to the extent of ascribing the responsibility for specific drug effects to certain atomic groups. Like Buchheim, he prioritized the experimental methods of physiology, at which he excelled thanks to his training in the laboratories of Carl Ludwig (1816–95), one of the pioneers of experimental physiology.[7]

However, Schmiedeberg also lacked interest in the details of drug action on the cellular level and consequently saw no necessity to develop a specific theory. Although it was known that the effects of a substance could be influenced by changes in its chemical composition, it was not clear at all to what extent chemistry offered a key to the understanding of drug action.[8] These limitations however were overshadowed by the general success of both researchers. After all, Buchheim and Schmiedeberg had turned pharmacology into a science.

Many of Schmiedeberg's students obtained positions as pharmacologists in Germany and abroad. Under his guidance countless new areas of pharmacology were opened up and investigated and many young scientists were trained in his laboratory in Strasbourg. In 1872, together with Edwin Klebs (1834–1913) and Bernhard Naunyn (1839–1925), Schmiedeberg founded the journal *Archive for Experimental Pathology and Pharmacology* (*Archiv für experimentelle Pathologie und Pharmakologie*), which was to represent the experimental paradigm.[9]

Buchheim's and Schmiedeberg's legacy lasted well into the new century, when many of their pupils were at their most productive. Among them were leading pharmacologists in Germany and Britain, such as Walther Straub and Arthur Robertson Cushny. John J. Abel, one of the most important pioneers of scientific pharmacology in the USA, was also educated by Schmiedeberg.[10] The main challenges though, were to concentrate on questions of drug binding on the cellular level, to perform research and to complement the findings of Buchheim and Schmiedberg and their students. Around the turn of the century it became increasingly important (or fashionable) to investigate the direct mode of action of drugs on cells and tissues. The morphologically-oriented medicine of the nineteenth century had brought only basic knowledge about the human body's physiology and pathology. Therefore, the most important aim of twentieth-century scientific medicine was the experimental investigation of disease *processes*, the application of physiology ('pathological physiology') to medical problems, and above all, the development of new therapies. Against the background of clinical needs it became more and more important to investigate the precise ways in which drugs influenced the organism, the organs, the tissues and especially the cells as the smallest units. The aim was to find specific cures for specific diseases and the development of new artificial drugs. The receptor concept responded to this demand. But for pharmacology it was a provocative theory – not least because it was developed by pharmacological 'outsiders'. From the perspective of pharmacologists three questions arose. First, should one adopt or at least consider the receptor concept? Or, second, should one reject it and present a new, competing concept? Or could one – as a third alternative – leave the Buchheim/Schmiedeberg research gap as it was, while taking a different route to increase the therapeutic efficiency of pharmacology? Straub, Cushny and many other pharmacologists favoured the second solution.

Walther Straub and Arthur Robertson Cushny: early twentieth-century pharmacology and the physical theory of drug action

(A)   Walther Straub and his 'poison-potential theory'

Walther Straub was one of the most important German pharmacologists of the first half of the twentieth century. Born in 1874, he studied medicine in Munich, Tübingen and Strasbourg between 1894 and 1897. He worked on his thesis 'On the conditions of the appearance of glycosuria after carbon monoxide poisoning'[11] in the laboratory of Oswald Schmiedeberg in Strasbourg. Then he became assistant to Rudolf Boehm (1844–1926) in the institute of pharmacology at the University of Leipzig and finished his habilitation thesis in 1905. In the same year he was appointed to the chair of pharmacology at the University of Marburg, and only one year later (1906) to the chair of the same field at the University of Würzburg. In 1907 he moved on to a similar position at the University of Freiburg. There he was the first full professor of pharmacology and became an important pioneer of his discipline. Between 1913 and 1917 a new institute was erected where Straub had a lot of talented assistants who later became professors of pharmacology at several German universities. Straub was so successful in Freiburg, that, in 1908, he was offered the prestigious chair of pharmacology at the University of Berlin, which still had a central role in German science. Tremendous efforts were made by the medical faculty in Freiburg as well as by the Duchy of Baden to keep Straub. They succeeded in the end and Straub stayed in Freiburg until 1923, when he moved to a chair at the University of Munich. He continued with his work in Munich, although it was overshadowed by financial strains and the delayed reconstruction of the pharmacological institute in 1932–33 as well as by the war. Straub died in Bad Tölz near Munich in 1944.[12]

Early in his career Straub gained an outstanding national and international reputation. In 1920 he founded the German Pharmacological Society (*Deutsche Pharmakologische Gesellschaft*), which was instrumental in institutionalizing the discipline.[13] From 1920 until his death he was the editor of the journal *Archiv für experimentelle Pathologie und Pharmakologie*.[14] In 1925 he became a member of the Deutsche Akademie der Naturforscher Leopoldina, in 1928, an honorary member of the American Society for Pharmacology and Experimental Therapeutics and in 1935, honorary member of the British Pharmacological Society.[15] In 1925, the conference of the German Pharmacological Society was held under the heading 'Straubismus convergens'.[16] In scientific discussions after the turn of the century Straub's voice had considerable impact on his colleagues. And this voice spoke strongly against the receptor concept.

Walther Straub had been influenced by the experimental approach of the Buchheim/Schmiedeberg school. As a keen animal experimenter, he developed new techniques to acquire knowledge about the functional aspects of drug action. One of the most important was the 'experimental system' (Rheinberger) that later became known as 'Straub's Frog Heart'; an isolated frog heart was fixed to a cannula and perfused with Ringer solution – a fluid containing sodium chloride, potassium chloride and calcium chloride to compensate dehydration – to examine active pharmacological substances. This 'experimental system' increased Straub's reputation and had a prolonged impact; indeed it is still in use in pharmacology. It was based on the heart preparation of physician and physiologist Oscar Langendorff (1853–1908), who developed the first method to investigate an isolated frog heart when perfusing it with blood or other nutritive substances. Therefore, Straub's innovation followed physiological tradition and reflected the transfer processes between the disciplines.[17]

Another of Straub's innovations was the so-called 'Mouse-tail-phenomenon' used to estimate the existence and amount of morphine in a drug dose. He also introduced the electrocardiograph (ECG) to pharmacological analysis, enabling him to record the activity of nerves and muscles, especially those of the heart muscle. Straub's ability to develop and apply new experimental techniques was so impressive that colleagues in the medical faculty of the University of Freiburg attended his lectures. Straub worked on the mode of action of many drugs, particularly on digitalis and 'luxury poisons', such as caffeine.[18]

For Straub, pharmacology had to be carried out by a specialist. It needed the help of chemistry, but was oriented mainly towards physiology as the guiding discipline. In 1920 he wrote:

> Today pharmacology is an experimental science which deals with the changes in states and processes in the living organism that are initiated by chemical and physical interventions ... It thus represents applied physiology and works with its methods, i.e. experiments in a physical and chemical direction. It represents experimental medicine and forms part of what is called pathological physiology.[19]

Straub tried to apply his expertise to contemporary problems in medicine. He favoured functional thinking, and he was a keen advocate of correlating pharmacological findings with clinical needs and purposes.[20] In this context he tried to investigate the way in which drugs act on living cells. Most important for his work in this area were two periods at the Zoological Research Institute in Naples in 1899 and 1905.[21] At this time, Straub was assistant to Rudolf Boehm (between 1898 and 1905) in Leipzig, and his work in Naples was devoted to research on the diffusion of alkaloids into the living cell. This work, especially that undertaken in 1905, laid the basis for future publications and for his theory of drug action.[22] Straub's preferred experimental animal organ was the heart of the sea snail, *Aplysia*.

Straub detected that muscarine had its effect on the heart of the sea snail only *while* it invaded the muscle cell. Only the *process* of invasion produced the effect. The amount of muscarine stored within the muscle of the heart had no impact on the strength of the effect; the substance was stored without chemical transformation.[23] Furthermore, the strength of the effect of muscarine was directly dependent on the concentration (the potential) of the poisonous solution. Straub saw the storage of muscarine as a process of increased inhibitory irritation (*Reiz*) of the heart muscle.[24] The constant application of the substance increased the periods of standstill of the heartbeat up to a maximum; thereafter, muscarine lost its effect. Straub concluded that an 'equilibrial reaction' (*Gleichgewichtsreaktion*) had taken place between the interior and the exterior milieu. This meant that muscarine had a reversible effect due to a difference of concentration between the inside of the heart muscle cells and the external environment, or due to the elimination of a certain potential of concentration in the direction of the interior of the cell.[25] While entering the cell, the muscarine stream would prevent the cell membrane functioning and block life-important chemical processes. But this process lasted only as long as the the cell had the capacity to store muscarine. Thereafter the inflow stopped and the equilibrium between the internal and external concentration of the substance ended any muscarine effect. Consequently, Straub generalized, 'that the processes in question seem to take place at the physical border of each individual cell'.[26]

These views corresponded to those of his teacher Rudolf Boehm, who had claimed in 1895 that the effect of a poison only increased up to a point of saturation ('*Sättigung*') of the cells, even if there was still poison in the blood.[27] Straub concluded that the effect of muscarine on the heart would not differ among different animal classes.[28] And he assumed that what would be true for muscarine would also be true for other alkaloids, for example, pilocarpine, physostigmine and nicotine, and for the hormone adrenalin.[29]

Therefore, it is not surprising that the research on alkaloids helped Straub to develop a general theory of the mode of action of drugs that opposed the idea of chemical drug binding. This became clear in the course of his work on drug antagonism. With explicit reference to the theories of Paul Ehrlich, Straub explained that the antagonism of muscarine and atropine did not follow the same laws as the toxin-antitoxin reaction. The latter relied on the saturation of chemical affinities. In contrast, the drug antagonism of muscarine and atropine relied on a fluent equilibrial process between the inside and outside of the cell, which was not dependent on chemical binding between the substances, nor was it subject to chemical laws.[30]

Straub's experiments also stood in contrast to John Newport Langley's theories of drug antagonism. Langley claimed that the poison would 'set a stimulus' to the cell in the case of chemical combination. After saturation, the cell received no more stimulation, and the effect of the poison correspondingly vanished. This explanation contradicted Straub's theory insofar as it assumed a chemical stimulus triggered by receptors and not a physical one triggered by a gradient of potentials. In Langley's view, atropine antagonized muscarine when it attached to the receptive substances and in this way prevented the latter's effect.[31]

Straub's work between 1899 and 1907, which comprised basic pharmacological research on the effects of drugs on the organism and its cells, led him to the development of a specific theory: the intensity of the effect of a poison or drug depends on the difference of its concentration between the outside and the inside of the cell. This theory was called the 'poison-potential theory' ('*Potentialgifttheorie*'). One also spoke of 'Straub's potential' ('*Straubsches Potential*'). In the following years in Freiburg, he elaborated on this theory,[32] describing it as a *physical* theory of drug action. A poison would find its way to the site of effect within the organism only with the help of its physical properties, its solubility in the wall of the cells, and its surface energy when dissolved. The chemical structure of poisons would not change when they affected an organism. All observations on the direct relation between structure and effect were speculative. As chemistry would not explain the effect of drugs in detail, it would not be possible to create artificial drugs on this basis. Ultimately, Straub – like Buchheim before him – presented a programme of 'experimental pharmacology' without explaining in detail any specific drug binding to cells.[33] He criticized Ehrlich's chemical side-chain theory and the related research on serum therapy, and later also the approach of John Newport Langley. As a keen defender of the discipline of pharmacology, to a certain extent Straub rejected Langley's approach as merely physiological. In Straub's view, Langley used chemical substances only to study the organism, while the pharmacologist would use the organism to study chemical substances.[34]

In 1910, Straub expressed doubts about the existence of receptors,[35] and in 1912 he took the last step in the development of his own theory, in a paper entitled 'The Importance of the Cell Membrane for the Effect of Chemical Substances on the Organism',[36] which explained the 'poison-potential theory' on the microscopical level. In his view, the specificity of certain organs for certain drugs could be explained only at the cellular level.[37] At this level, however, Straub was forced to discuss Ehrlich's chemical theory of drug binding. In 1902, Ehrlich had criticized pharmacology for having neglected research on the relationship between chemical constitution, distribution and pharmacological effect,[38] so it is not surprising that Straub launched a counter-attack against Ehrlich's side-chain theory. The reaction of a substance with the cell on the basis of chemical binding went 'too far as a general approach' and was 'not admissible and finally not fruitful'.[39] There were countless substances which simply had no ability to react chemically with the organism, for example, indifferent narcotics. Straub concluded that by assuming the existence of chemoreceptors 'one miracle is changed into another. The existence of chemoreceptors for poisons should not be denied, but it is not general and it is not possible to base a far-reaching theory on this finding.'[40]

Straub was reacting to Ehrlich's Nobel Prize lecture, published in 1909, on the 'partial functions of the cell', in which Ehrlich had described Straub's understanding of the side-chain theory as false and simplistic.[41] In Straub's view, more basic research was needed before going on to speculate about receptors.[42] For Straub, research on the cell membrane was important because he believed that its function was ruled by physical laws.[43] In this sense Straub's thinking corresponded to the contemporary trend of broadening the scope of research on the cell, extending nineteenth-century analysis of the internal structure of the cell to cell-to-cell interaction or exchanges via cell membranes.[44] However, Straub remained sceptical as problems with the intricate physical and physico-chemical processes in cell membranes persisted. This meant that it would be impossible to predict which substances would be able to pass into cells. The only successful path for producing new drugs would be by empirical testing.[45]

(B) The recognition of Straub's theory of drug action

With his 'poison-potential theory', which was completed in 1912, Straub became part of a large group of 'physical theorists' concerned with drug binding, a subject that dominated the international pharmacological scientific community between about 1900 and 1945. This group opposed the chemical theory of drug binding, and sub-groups based their arguments on those drugs which served best their respective hypotheses. Straub's approach itself amalgamated the results of his early investigations undertaken in Naples with those of other physical theorists. These were mainly the studies related to the colloidal theory of Karl Landsteiner (well known in 1912, when Straub launched his attacks on Ehrlich) and to the theories of Meyer and Overton (see below).[46]

There were considerable overlaps with the research of other scientists, and Straub's theory had many supporters, explaining, as it did, in a 'simple manner the transient effects produced by many drugs'.[47] But Straub's approach did not remain unquestioned. One point of criticism was the fact that according to Straub's theory of potentiality drugs could only show transient effects. Others argued that some of Straub's results had been achieved as a result of faulty experimental conditions.[48]

Although there was severe criticism of Straub's theory, especially in the 1930s (see Chapter 5 below), Straub did not give up.[49] From the time he developed his theory until the end of his life, he tried to improve it using diverse pharmacological investigations. In 1916, for example, he tried to estimate the most effective dose of different sub-groups of digitalis-related compounds. On the one hand, Straub explained that the different effects of the substances depended on their different chemical configurations, on the other hand, he refused to accept a chemical theory of drug binding because 'speculations on the *immediate* relations between constitution and effect and the character of the binding of poison and living organ are as useless as ever'.[50] Thirteen years later, in his Lane Lectures presented in San Francisco in 1929, he conceded 'that an invisible process of a chemical nature must take place, a process of anchorage, between the contact of the poison with the cardiac muscle cell and the appearance of visible action'.[51] On the other hand he postulated that the binding of the poison was 'most probably an adsorption process'. Chemical processes 'may take part. However, these latter processes are unknown...' Hinting at his poison-potential theory, Straub explained that extensive chemical processes in the cell need not occur, 'for when a cell membrane has become so impermeable that exchanges are impossible the cell must die'.[52] Also, in the case of alkaloids, Straub insisted on the impact of concentration potentials and the balance between the inside and outside of the organ or the cell.[53] Finally, in his Lane Lecture on heavy metals, he claimed 'that the reaction of a metallic ion with serum is a reversible equilibrium-reaction'.[54] In 1936, Straub wanted to demonstrate his poison-potential theory when repeating Otto Loewi's (1873–1961) investigation of the humoral transfer of the vagus effect on the heart, but he was not able to increase the experimental evidence in favour of his own theory.[55]

It is worth noting that, in spite of all his efforts to implement a new theory of drug action, Straub did not need this theory to achieve new results in his research areas. The bulk of his investigations relied on the traditional Buchheim-Schmiedeberg approach of applying substances to animals or their organs and watching, measuring and evaluating the physiological reactions of the organisms. The specific mechanism of drug binding played no major role in this experimental setting. In 1911, he injected lead beneath the skin of cats and observed the outcome, including the excretion of lead via the kidneys and bowels. During the First World War, Straub used the same method on magnesium sulphate applications in cases of tetanus, and on the use of antitoxin against gas gangrene. In 1940 he joined with a colleague to publish a study on the effect of nicotine and the dietetics of smoking; they injected cats with nicotine and examined how they reacted. He then went on to draw conclusions about the effect of nicotine on human beings.[56]

After 1912 Straub wrote only one paper devoted explicitly to the discussion of the chemical and physical theory of drug action. It was a short comment on Alfred Joseph Clark's 1936 attacks on Straub's theory (see Chapter 5).[57] Therefore, one has to assume that Straub's motivation in formulating the poison-potential theory was primarily a response to then 'fashionable' scientific conventions. In the early decades of the twentieth century, it was not acceptable to ignore the problem of the intricate mode of action of drugs on organs and cells. There were many uncertainties concerning the application of chemistry to medicine and biology, and Paul Ehrlich's side-chain and receptor theory was hotly debated. In this context Straub developed his theory in line with accepted applications of physical theories for explaining the functions of the human body.

(C) Arthur Robertson Cushny and the receptors

Straub was not the only leading pharmacologist who was a devoted student of Buchheim and Schmiedeberg as well as an adherent to the physical theory of drug action. Sir Arthur Robertson Cushny (1866–1926) played the same role in Britain in this regard as Straub did in Germany.

Cushny had studied medicine at the University of Aberdeen between 1886 and 1889. There he met one of the few experimental pharmacologists in Britain at this time, John Theodore Cash (1854–1936), who persuaded him to go to Schmiedeberg for his further education. After having spent some years in Berne and Würzburg, Cushny headed for Strasbourg and became an assistant to Schmiedeberg (1892–93). He had planned to go back to Britain to combine pharmacological laboratory studies with clinical work in a hospital, but instead he accepted an offer from the founding father of American pharmacology, John Jacob Abel (1857–1938) to succeed him in 1893 in the chair of pharmacology at the University of Ann Arbor in Michigan. There he spent twelve years, before moving to London in 1905, where he was appointed to the newly founded chair of pharmacology at University College. In London, he built up the department of pharmacology and thirteen years later, in 1918, he was appointed to the chair of pharmacology at the University of Edinburgh. Once again he performed pioneering work, introducing modern experimental pharmacology. He filled this post until his sudden death in 1926.[58]

Cushny was the founding father of pharmacology in Britain, and he had considerable impact on the development of the field not only at home but also in the USA. Several pharmacology chair-holders in England in the 1920s had been assistants of Cushny.[59] He worked along the lines of Buchheim and Schmiedeberg, and like them he was devoted to experimental pharmacology and tried to transform drug therapy into an effective tool of clinical medicine.[60]

His concern with practice made him reluctant to start a career as a university lecturer after his medical education and it also influenced his move from London to Edinburgh, where there seemed to be better links with practical therapeutics.[61] In 1899, Cushny wrote the first pharmacological textbook in the English language, which was oriented towards Schmiedeberg's pioneering work *Grundriss der Arzneimittellehre* (Outline of Pharmacology) and dedicated to his teacher. The subtitle of his book expressed Cushny's strong bonds with the science of physiology as it dealt with 'The action of drugs in health and disease'.[62] Cushny worked on 'the development of physiological research in its relations to pharmacology', and in 1923 he was one of the organizers of the International Physiological Congress in Edinburgh.[63] Like Straub, Cushny saw himself as a pioneer of an experimental pharmacology with therapeutic impact. He too broke new ground in a large number of different areas, such as the pharmacology of digitalis, the therapeutics of heart disease, urinary secretion and the function of the kidneys.[64] As with Straub, it was this interest in the direct mode of action of drugs which led him to deal with the newly developed receptor concept. He found that the response of tissues of higher animals to certain alkaloids and bases depended on the alkaloids' constitution in terms of optical isomers.[65] Cushny believed it possible to correlate different effects of substances on organs with their isomeric constitution.[66] From 1903, Cushny published on this topic, and it remained a pet subject until the end of his life.[67] Again like Straub, Cushny had developed a physical theory of drug action which was potentially able to explain basic phenomena of life. In 1909, he wrote to Abel: 'Optical activity interests me very much. It is the one sign of living matter that we have, it seems to me, but it is still so obscure physically that not much is to be done with it.'[68] In Cushny's view, optical activity was 'the most persistent evidence of life which we possess', and it was a physical quality of the substances that determined their mode of action in the body.[69]

The chemical structure of drugs was accepted by Cushny as the underlying basis of pharmacology and isomeric activity, but he saw the effect of drugs as correlated with differences in isomeric activity and not just with differences in chemical structure. Physical qualities, including volatility and solubility, played a greater role than chemical characteristics. He believed in 'receptive substances' of the cell (1908), which could combine with the two different isomers of a substance, turning the light left or right (l- and d-isomers), although he could say almost nothing specific about the nature of the binding and of the affinities between cells and drugs on the basis of isomeric activity. As with Straub, his experimental approach did not allow results in this direction. Cushny could only offer a general hypothesis, namely that receptors have a general affinity to optically active substances because they themselves are living matter. For him, specific isomeric differences of two substances had nothing to do with their binding capacities to receptors.[70] For Cushny, the receptors were merely tools of the cells, used to organize unspecific chemical binding, but the effect of the binding was decided by basic physical qualities, by the character of the molecule as a whole.[71]

Therefore it is not surprising that Cushny, like Straub, opposed Ehrlich's and Langley's theories of chemical drug binding. He rejected Ehrlich's theory that the side-chain of a therapeutic substance would have greater affinity for the parasite than for the cells of the host. Cushny believed that the drug was able to combine with the microbe more *quickly* than with the bodily cells. But Cushny also had general objections against Ehrlich's theory. Compared with its experimental basis, the conclusions were too far-reaching and too speculative: 'I had to look up a good deal of the toxin-antitoxin literature, and I was rather dismayed at the indefiniteness of much of the experimental results on which so much theory has been raised.'[72]

According to Parascandola, Cushny developed his theory of isomers gradually and he never gave it up. Parascandola concedes however that by 1925 Cushny would have been more aware of the difficulties in differentiating between the physical and chemical qualities of drug binding.[73] Cushny was indeed a defender of the physical theory but the course of his opinions was more complex than Parascandola allows. If we examine the different editions of Cushny's textbook, it is clear that at the beginning of his academic career, in the first and second editions (1899, 1901), he acknowledged the chemical theory as a model which had to be taken into serious consideration. Although he remarked that 'the relation between chemical constitution and pharmacological action can be followed only a short distance as yet', he also wrote that 'the great majority of drugs act through their chemical affinity for certain forms of living matter. They probably form temporary combinations with some forms of protoplasm, and alter the function of all cells which contain these forms.'[74]

During the next ten years Cushny developed a more critical attitude towards the chemical theory of drug attachment. In the third and fourth editions of his textbook (1903, 1907), he drew the attention of the reader to new developments in the field of physical chemistry.[75] He pointed out that this new direction of research attributed the effects of drugs mainly to the 'physical structure of the living cell rather than to its chemical constitution'.[76] This physicalist tendency became even stronger in 1910, when in the fifth edition he integrated the new research strands.[77] Although he avoided presenting a one-sided description of recent research in his field, and although he explicitly acknowledged the work of Ehrlich and Langley, Cushny's preference for the physical theory was clear. His chapter on the 'Mode of Action of Drugs. Stimulation, Depression and Irritation' dealt extensively with the physical theory, and Walther Straub's approach was mentioned in this context. In the chapter on 'The Relation between Chemical Composition and Pharmacological Action', Cushny remarked: 'As a matter of fact the physical properties of drugs appear to have a more direct bearing upon their action than the chemical structure; that is, the properties of the molecule as a whole determine its effects more than any of its constituent parts.'[78]

The sixth edition of Cushny's textbook (1915) again shows a certain change of direction. The chapters on the mode of action of drugs and on the relation between chemical composition and pharmacological action were combined under the heading 'General Theories of Pharmacological Action'. The title as well as the contents now underlined Cushny's intermediate standpoint in a much more precise manner; he had not had the breakthrough for which the physicalists had hoped.[79]

During the period 1910–15, in spite of his own more or less physical theory of drug action, Cushny maintained a moderate position. In the seventeenth edition (1918) and in the eighteenth and last edition, which was still edited by him in 1924, he made no changes in the chapter mentioned above. However, the 1924 edition, for the first time, included the term 'receptor' in the index.[80]

The analysis of Cushny's textbook shows another similarity with Walther Straub. Both had to make increasing concessions to the chemical theory. That this happened only gradually is partly due to Cushny's motto – not to abandon a theory merely because of some criticism, but to revise it carefully on the basis of further examination and feedback from the critics.[81] Cushny's development illustrates the way in which the chemical approach gained more and more ground even among its critics. This development helped to pave the way for researchers such as Alfred Joseph Clark, who embraced the initiatives of Ehrlich and Langley more fully.[82]

Other representatives of the 'physical theory of drug binding' and its general acceptance in pharmacology

The work of Straub and Cushny illustrates the strong influence of the physical theory on early twentieth-century pharmacology. There were many proponents of a physical theory of drug action within and without pharmacology and it was well represented in contemporary textbooks. Within pharmacology the proponents of the physical theory were the largest group and it was the accepted theory of the discipline. Chemical theorists found themselves exceptions to the rule. This made it difficult for supporters of the chemical theory, such as Langley and Ehrlich, to achieve a breakthrough for their theories in that field. The good relationship between Straub and Cushny, even in the difficult times of the First World War when links between scientists from Germany and other Western countries were severely disturbed, shows how closed the ranks of the physical theorists were.[83] We will now discuss the network of physical theorists, the variety of other important contemporary physical theories, their influence on pharmacology and the challenge that they posed to the chemical theory.

(A) Hans Horst Meyer and Charles Ernest Overton

An area of study that had considerable influence at the beginning of the twentieth century was research on the effects of narcotics undertaken by the pharmacologist Hans Horst Meyer (1853–1939). He was appointed professor in Marburg in 1884 and in Vienna in 1904; he was (like Straub) a pupil of Schmiedeberg and also of the Zurich biologist and pharmacologist Charles Ernest Overton (1865–1933). Meyer and Overton based their concept on the physical quality of solubility, claiming that substances passing through the semi-permeable cell membrane were more soluble in fatty oils and lipoids than in water. Since it was assumed that only fat-soluble substances passed through semi-permeable membranes, it was concluded that the membranes must consist of fats or fat-like substances, such as lipoids and cholesterol. Those substances soluble in fats had a narcotic effect and their effect was particularly strong on nerve cells as the chemical structure of the latter depended on lipoids.[84] Although Meyer and Overton acknowledged the chemical properties of the substances and of the respective cell membrane, they claimed that physical qualities, such as adsorption, were decisive for the effect of the drugs. Decisive for both researchers was the efficacy of those narcotics. Meyer especially saw his contribution to the research on attachment of drugs to tissues as successor to the work of Buchheim and Schmiedeberg.[85] Meyer and Overton had considerable influence on contemporary pharmacologists, among them Straub and Cushny, the latter being a close friend of Meyer's.[86]

(B) Isidor Traube

Another advocate of the physical point of view was the physiologist and biochemist Isidor Traube (1860–1943), who worked at the Technische Hochschule in Berlin-Charlottenburg. In 1919, he published a paper on the physical theory of the action of drugs and poisons. The physical characteristics of substances that in Traube's view were decisive for drug binding were surface activity, solubility, adsorption, osmotic power, friction and dispersal; their pouring, flaking and catalytic properties; and their electric potential and electric power (*Ladung*).[87] In his explanations, Traube focused strongly on the surface energy of substances, that is, their physical potential to disperse through the cell membrane. The most important role in Traube's theory was played by the 'attaching pressure' between substance and cell. His theory was accordingly called the 'Theory of Attaching Pressure' ('*Haftdrucktheorie*'). The lower the 'attaching pressure' (*Haftdruck*), the higher the surface activity of a substance.

And the more surface activity a substance had, the greater was its capability to flake and to invade the water-soluble cell-protoplasm. This meant that it was important to measure the surface activity of substances, especially of those substances that produced an effect in very small doses.[88]

Like Meyer and Overton, Traube illustrated his theory with the example of narcotics, but also with other drugs, such as the stimulant camphor or with disinfecting substances.[89] Remarkably, he and his students also detected the importance of the surface activity in alkaloids. As there was, in his view, only poor knowledge about the direct effects of drugs and their site of action,[90] Traube criticized Ehrlich's chemical side-chain theory. He supported his approach using Landsteiner's colloidal theory of drug action, which also relied on the propagation of physical forces as the basis of drug effects. In Traube's view, chemotherapeutic research should develop into 'physico-therapy' (*'Physiko-Therapie'*), concentrating on the investigation of the physical properties of drugs.[91]

Colloidal chemistry lay behind the work of all the physical theorists, including Straub. It occupied the borderland between physics and chemistry and appeared to explain drug binding to cells far better than Ehrlich's and Langley's receptor concepts. Its reputation grew from the beginning of the twentieth century and Ehrlich had to defend his receptor theory against one of the most prominent supporters of colloidal chemistry, Karl Landsteiner (see Chapter 1 above). In the years after Ehrlich's death, colloidal chemistry, which 'describes the physico-chemical behaviour of particles in a solvent or at a boundary between solvents',[92] became even more fashionable. Colloids consist of large particles that do not dissolve when brought into contact with solutions; rather, they form gels. These gels form a layer around cells. The surface area between solvents and cells became decisive for explaining exchange and binding processes. Surface tension, surface activity and adsorption as mechanical as well as electrical phenomena played an important role in these processes. Colloidal chemistry cut across all purely chemical explanations and invaded the territory of physics. It was fundamental to the understanding of life as such: 'Since colloidal behaviour was typical of proteins, and proteins were typical of life, colloid chemistry seemed to offer a new key to the physics and chemistry of life.'[93] But although colloidal chemistry supported the reception of physical theories of drug action, it is important to note that the integration of chemistry meant that the door was not shut entirely to chemical theories of drug binding. And bearing in mind that Straub and Cushny were also forced to make more and more concessions to the latter, it was then easier for

Alfred Joseph Clark to present his quantitative receptor concept in the late 1920s and the 1930s (see Chapter 5).

Meyer and Overton's approach, the attachment-pressure theory of Traube and the colloidal theory were all widely applied. For example, William Maddock Bayliss (1860–1924), professor of general physiology at University College London, in 1915 favoured the physical theory on the basis of colloidal phenomena and the related process of osmosis. Bayliss mainly attacked the claim to exclusiveness of the chemical theory, arguing that Paul Ehrlich's receptor theory was far too dogmatic.[94] Bayliss was acknowledged by the pharmacologists of his time: he had good contacts with Walther Straub, and the son of Rudolf Boehm had worked in Bayliss's laboratories.[95]

(C) Walter Ernest Dixon

The pharmacologists' instrumentalization of the physical theories as well as the slowly growing influence of the chemical approach can be illustrated by the work of Walter Ernest Dixon. He proposed no original theory of drug binding; rather, his work reflects both the broad impact of the physical theories on many contemporary pharmacologists and the gradual general acceptance of chemical thinking and the chemical theory. Like Cushny, Dixon was a pioneer of pharmacology in Britain. Following his appointment as a lecturer (1909), he became in 1919 the first reader in pharmacology at Cambridge University. For a time he held concurrently the chair of materia medica at King's College London. He was prominently involved in establishing pharmacology as an independent medical discipline in Britain and in building up its reputation as a practice-oriented field that supported clinical medicine.[96] In accordance with the general trend in pharmacology after the turn of the century to explore the mode of action of drugs on cells, in 1909 Dixon and his colleague Philip Hamill performed animal experiments on the effects of strychnine on the spinal cord and of secretin on the pancreas (see also Chapter 2 above). These investigations were stimulated by the receptor theories of Ehrlich and Langley; Dixon and his colleague wanted to test the 'validity of this hypothesis'.[97] Their work lent 'no support to the chemo-receptor hypothesis' as far as the effect of alkaloids (such as strychnine) was concerned.[98] Dixon applied Ehrlich's methods of investigating the effects of toxins, for example, mixing tissues and substances to observe a proposed decrease of activity of the substance in question. However, even this paper showed Dixon as a representative of the old Buchheim/Schmiedeberg approach to animal research, which focused on the observation of physiological effects of drugs on organs and tissues without following up the question of drug binding in detail. Dixon was not able to find any evidence for the attachment of alkaloids to specific cells but speculated that poisons released hormones in the body that then bound to cell receptors.[99]

Early in his career, Dixon was a supporter of the physical theory of drug action. Nonetheless, in the first edition of his textbook *A Manual of Pharmacology*, published in 1906, he conceded a strong role to chemistry in the attachment of drugs to cells.[100] 'Most drugs', he wrote, 'exert their action by chemical, and not physical, means.' Furthermore there was 'some sort of combination with a chemical body contained in the cell acted upon; and it is generally assumed that this combination is chemical in character'. But the 'constituent of the living cell' that the drug interacted with, was 'generally unknown, and can rarely be subjected to chemical analysis'. Furthermore, it would not be possible to explain the pharmacological action of a drug through its chemical constitution. The change of action of a substance after having altered its chemical constitution was often caused mainly by physical changes (for example, ionization, absorption, resorption). In Dixon's view and according to Ehrlich's own views before 1907, Ehrlich's side-chain theory was not applicable to drug binding because drugs would not bind firmly to tissues and often could be removed easily. In general, there was a clear emphasis on the physical theory.[101]

Dixon's description did not change in the second edition of 1908.[102] In the third edition, published in 1913, Dixon refused to use the terms 'receptor' and 'myoneural junction', and this was pointed out explicitly in the preface: receptors and myoneural junctions were 'only words without precise meanings and cannot fail to present difficulties to the student without in any way assisting him'.[103] In the same edition, Dixon's earlier arguments in favour of a chemical interpretation were further weakened: whereas in the first two editions it was 'certain' that when the drug produced an effect there would be 'some sort of combination with a chemical body contained in the cell acted upon', it was now only 'suggested'. Also he strengthened the physical aspect by inserting comments on the physical theory of Isidor Traube, in addition to those on the theory of Hans Horst Meyer that he had already referred to in the first two editions.[104]

It was only in the fourth edition, in 1915, that Dixon was forced to rewrite his textbook because of 'the increase of knowledge of the mode of action of drugs which has been obtained during the last few years'.[105] But he did not change his basic attitudes and the main parts of the book dealing with the physical and chemical properties of drugs and their effects remained substantially the same. On the other hand, he added a discussion of Walther Straub's approach and Meyer, Traube and Straub were described as supplying 'a useful hypothesis' representing 'a step forward towards a correct understanding' of drug action.[106] But at least Dixon considered – nine years after its first publication – Thomas Renton Elliott's suggestion of a 'myoneural junction' between nerve-ending and end organ (see Chapter 2 above). Since substances such as atropine or ergotoxin could influence the action of nerves without affecting the cells of the targeted organ, Dixon had to concede that 'it becomes necessary ... to introduce a new structure, neither an integral part of the nerve nor end-cell, which we may call for the time the myo-neural substance'.[107]

Dixon maintained his views, although chemical explanations of drug binding were given more room in his textbook. In the fifth edition (1921) he inserted a chapter on 'Chemotherapy'. There he described Ehrlich's concept of the '*Therapia sterilisans magna*', that is, of finding an agent causing harm only to microbes but not to the host, and the side-chain theory as its basis. But in Dixon's view there was 'no valid evidence ... for such a conception', and he argued that the cells of the host took a much more prominent role in combating infectious diseases than the combination of the drug with the side-chains of the parasite. Thus chemotherapy was for Dixon 'a speculation based neither on chemistry nor pharmacology'.[108]

In the sixth edition (1925), the problem of the specific mode of action of drugs was again dealt with in the preface. With persistent reluctance to accept the chemical theory's explanation for the attachment of drugs to cells, Dixon tried to find a compromise:

> certain it is that drug action is not determined directly by chemical combination with body constituents, but rather by delicate physical processes such as those of absorption, solution, and surface tension. On the other hand, slight alteration of a molecule already complicated and with a known action has led to the production of many useful compounds, and not infrequently we may foresee the type of action which will occur under such special conditions.[109]

In the chapter on 'Chemotherapy', he acknowledged Ehrlich's efforts in finding specific drugs against microbes, writing that the 'latest member of this class, the so-called "205", promises, however, to be an invaluable treatment of trypanosomiasis'. Dixon's evaluation of chemotherapy was much less critical than four years earlier as he pointed out its relevance as a specific therapeutic method.[110] The seventh edition – the last to appear in Dixon's lifetime – was published in 1929 with the same preface and without substantial changes.[111]

That the chemical theory of the attachment of drugs to cells maintained its influence can be demonstrated by those few scientists who, in the face of strong criticism and in spite of the uptake of colloidal chemistry, remained its loyal supporters. William Whitla, professor of materia medica and therapeutics at Queen's University, Belfast, maintained in his textbook – well up to the third decade of the twentieth century – that the selective power of drugs over particular tissues and organs was 'purely chemical'. After 1915, he praised Ehrlich's Salvarsan and the immunological work of Almroth Wright, pathologist and immunologist at St Mary's Hospital, London, as positive examples of the practical application of the chemical theory of drug action, and in 1923 Whitla used the term 'receptors'.[112] Furthermore, those pharmacologists who from the beginning held an intermediate position between the two camps are also worthy of comment.

(D) Torald Hermann Sollmann

The work of Torald Hermann Sollmann (1874–1965), first assistant professor and later professor of pharmacology and materia medica in the school of medicine of Western Reserve University, Cleveland, USA, offers an example of gradual change from support of the physical theory to support of the chemical theory.[113] Sollmann was an experimental pharmacologist who at the beginning of the twentieth century not only wanted to investigate the 'physiologic action' of drugs, but beyond this 'the reasons for these actions'.[114] As early as 1901, Sollmann acknowledged the chemical nature of the drug-cell contact and of drug effects when he stated that with the introduction of 'a strange molecule' into the cell's protoplasm 'things go entirely different'.[115] Pharmacological action 'must be conceived as purely chemic'. Sollmann pointed out that it was frequently the case that substances of similar structure had a similar action, and that 'similar structures are affected by the same drug'. Although this supported the chemical theory, there were 'many factors here which we do not understand'. Sollmann's argument was that drugs with the same 'elementary composition' might have a different constitution and this would be the case with isomeric compounds – the central tenet of Arthur Cushny's physical theory of drug action. Another problem that worried Sollman was that 'identical actions may be obtained from substances having a totally different chemic character'. In his view Ehrlich's side-chain theory rested 'entirely upon speculative grounds'.[116]

In 1922, Sollmann reiterated the essence of his earlier remarks, namely that it was not possible to make a clear statement about the role of chemical and physical laws in the attachment of drugs to cells:

The processes of life are essentially conditioned on chemical and physical changes in the constituents of the cells ... Our limited knowledge of the chemical details of the living cell does not permit any deep insight into the nature of the action of these substances, except in a few directions. They suffice to show that the mechanism of the action of different drugs is not uniform, but is sometimes along chemical, and sometimes along physical lines. No sharp division can be drawn.[117]

For Sollmann, it was impossible to 'construct a general theory of pharmacological action on chemical lines', but he saw the chemical theory as a useful aid in explaining those gaps which the physical theory had left: if 'a chemic substance possesses actions for which there is no adequate physical explanation, we presume that it enters into chemic reactions with the protoplasm'.[118]

Sollmann continued to support the 'chemic theory'. In 1928, he even elaborated on it. Together with Paul John Hanzlik (1885–1951), professor of pharmacology at Stanford University, San Francisco, he published *An Introduction to Experimental Pharmacology*, in which the two researchers presented a whole section on what they called the 'receptive mechanism'. Remarkably, their results were based on animal experiments on the eye that were very much like those of Langley 22 years earlier (see Chapter 2 above). The paralysing effect of atropine as an antagonist of pilocarpine on the sphincter muscle of the pupil, even after degeneration of the relevant nerve, would be caused by: 'something between the neurone and the muscle-contractility; something which may be conceived as "receiving" the stimulus of the nerve and of the pilocarpine and which may be called the "receptive mechanism", to avoid premature adherence to rigid theories.'[119]

Sollmann and Hanzlik, with explicit reference to Langley's monograph of 1921, were reluctant to accept the term 'receptive substance' or 'receptor' and other related terms, as these would 'imply more than is really known'. The authors left the question open as to whether there was 'a specialized structure of the muscular protoplasm', a 'specially excitable substance', or a 'specially labile "side-chain" of the contractile substance'. In their view, there was no experimental evidence for such suggestions, but as it seemed to them that there was something between nerve ending and muscle, they concentrated on the functional aspects and described the whole process with their new term, 'receptive mechanism'. Sollmann and Hanzlik's notion of a receptive mechanism was the first new concept directly derived from Ehrlich's and Langley's ideas about receptors.[120] By the end of the 1920s, the way had been prepared for a fuller acknowledgement of the receptor idea in pharmacology, although the scientific community of 'physical theorists' had refused to accept it for nearly two decades.

Other theories of drug action in the first three decades of the twentieth century

The previous sections have concentrated on the prevalence of the physical theory in pharmacology at the end of the nineteenth century and in the early years of the twentieth century. In this section we discuss some other prominent, though somewhat 'eccentric' theories of drug binding whose existence illustrates the difficulties and confusion of the period in explaining how drugs acted on cells. It was hard for any one concept to claim superiority over its competitors.

(A)   The Weber-Fechner Law

One important theoretical approach to the explanation of drug action was the so-called 'Weber-Fechner Law', which was based on scientific developments in nineteenth-century Germany. Gustav Theodor Fechner (1801–87) was a philosopher, physicist and psychophysicist at the University of Leipzig between 1834 and 1887. He was primarily concerned with philosophical problems that were strongly influenced by early nineteenth-century Romanticism. Fechner reflected on the 'body-soul problem', with the aim of uncovering the relations between the two. Against this background he developed 'psychophysics' as the science of the interrelations between the bodily and the psychic world. The development of his theories was very much influenced by discussions among a Leipzig circle of scientists and upper middle-class university teachers. One of them, the physiologist Ernst Heinrich Weber (1795–1878), had postulated the 'Weber Law', namely that in test series of clearly differentiated levels of sensations the sensibility of the body increases proportionally to the increase of the stimulus. On the basis of Weber's Law, Fechner carried out calculations, for example, geometrical rows of numbers, and developed a second law, namely that sensation was proportional to the logarithm of the stimulus ($E = \log R$). This was called the 'Weber-Fechner Law'.[121] Because the sensory stimuli were increased with a constant relationship over time, the results of the experiments related to the Weber-Fechner Law showed 'a linear relation between the amount of response of a sensory organ and the logarithm of the intensity of the stimulus'.[122]

The 'Weber-Fechner Law' was one of the most influential scientific theories of the nineteenth century.[123] It was intensely debated from the time of its inception, and debate continued up to the Second World War. These debates were mainly related to research in physiology, including John Newport Langley's field of interest.

The application of the Weber-Fechner Law in pharmacology relied on basic research in physiology. In the physiological laboratories at the University of Cambridge, Langley's workplace, Bryan H.C. Matthews (1906–86) was Beit Memorial Fellow of King's College Cambridge and worked in the 1930s on the response of muscles to stimuli. In 1931, Matthews published a paper on this topic.[124] He made a preparation of a single muscle, namely the toe-muscle of the frog. The reaction of the muscle to each stimulus applied was a single rhythmic contraction. This response, in Matthews's view, represented one single end organ being one single nerve ending.[125] He found that the frequency of response was roughly proportional to the logarithm of the electric charge. With this result, Matthews confirmed the Weber-Fechner Law, which he explicitly mentioned. Matthews used the term 'receptor' at the beginning of his paper in connection with the transmission of nerval impulses, but it played no important role on the following pages. In the summary of his study, the word 'receptor' did not appear, as the most important facts concerned the reaction of the muscle fibre – the end organ – on specific irritations. Matthews's paper was very much indebted to the work of his teacher and head of department, Edgar Douglas (Lord) Adrian (1889–1997), who had collaborated on a paper in 1926, which presented the outcome that all end organs would react qualitatively in the same way when giving rhythmic discharges. This meant that the investigation of any type of end organ would give results which would be generally applicable.[126]

Other physiologists argued against the general applicability of the Weber-Fechner Law. August Pütter (1879–1929), who in 1923 became professor of physiology in Heidelberg,[127] followed the strand of physical chemistry, studying the metabolism of irritated organs. He ended up with different results, thus becoming a critic of Weber and Fechner, but, in principle, his theory rested on the same basis, that is, the application of mathematics to sensations and reactions.

Pütter's experiments nonetheless had an impact on pharmacology. In a paper in 1918, Pütter was mainly interested in investigating the metabolism and exchange of substances (*Stoffumsatz, Stoffaustausch*) and the relations of these two processes to the phenomena of irritation.[128] In his view, the metabolism of the body or organs in cases of irritation was ruled by physical as well as chemical laws, and it relied on so-called 'sensible agents' ('S-agents'). In the course of the process of irritation, S-agents would be transformed into 'irritating agents' (*Erregungsstoffe*, 'R-agents'). This meant that the concentration of R-agents determined the possibility of irritation and the condition of irritation. Decisive for Pütter's theory was that irritation happened only if the R-agents had reached a certain threshold – a basic or 'zero-threshold' in case of a new irritation (*Nullschwelle*) or a 'difference-threshold' (*Unterschiedsschwelle*) in case of an intensified irritation. The time necessary for the creation of the threshold-irritation was an exponential function of the intensity of the irritation. And vice versa: the intensity of irritation necessary to create a threshold-irritation was an exponential function of the period of time of irritation. In this way Pütter criticized the Weber-Fechner Law, which claimed a fixed relation between degree of irritation and sensation: only at the beginning did the degree of irritation and sensation correspond. After some time the degree of sensation could no longer be increased by prolonged stimulation.[129]

Selig Hecht (1892–1947), professor of biophysics at Columbia University in New York, dealt with the Weber-Fechner Law in a way that resembled Pütter's.[130] He was interested in the relationship between sensation and irritation in the case of vision. In 1931, he wrote a textbook chapter on the physical chemistry and the physiology of vision.[131] On the basis of a short history of the Weber-Fechner Law, Hecht argued that the process of vision rested on photochemical (physicochemical) reactions and that there was a threshold for irritation. Hecht himself provided evidence against the Weber-Fechner Law with experiments on the sensitive organs of the oyster, *Mya arenaria*. The error of the psychophysical law was, in Hecht's view, that it ignored one main characteristic of the registration of light intensity, namely discontinuity. Hecht quoted other authors who showed the inconstancy of the relationship between sensation and irritation, thereby refuting Weber-Fechner. Remarkably, Hecht wrote of 'receptors' that were responsible for the uptake of impulses that thereafter went to the retina. Hecht also quoted Langley, who had shown that nerve impulses were based on preformed structural configurations.[132] But Hecht's description of the receptor remained unclear because he used the term in connection with the analysis of processes as well as in the explanation of morphological structures.

The Weber-Fechner Law had serious problems achieving acceptance within the scientific community of pharmacologists as an explanatory tool for drug binding: the constant relationship between stimulus and effect was refuted in many experiments, and the law was found inadequate as an explanation of either drug action or the effects on receptors.[133]

(B)   The Arndt-Schulz Law and the impact of homoeopathy

An interesting interpretation of drug action was the so-called 'Arndt-Schulz Law'. Compared to the Weber-Fechner Law, it only had a minor impact on pharmacological research, but its origins also lay in nineteenth-century Germany and its acceptance was problematic from the start. The Arndt-Schulz Law stemmed from the theories of the Greifswald psychiatrist Rudolf Arndt (1835–1900), which had been first put forward in 1885. Arndt had suggested 'that if a weak stimulus excites an organism, then any drug in sufficiently weak dose ought to do this as well'.[134] This suggestion, which was influenced by homoeopathic thinking, was developed in cooperation with Hugo Schulz (1853–1932), who since 1883 had held the chair of pharmacology at the University of Greifswald. Schulz never regarded himself as a homoeopath, but he clearly spread related ideas in his lectures. The general form of the Arndt-Schulz Law was that weak stimuli cause the emergence of vital processes, medium stimuli partially support them, strong stimuli inhibit them, and very strong stimuli produce complete inhibition.[135] In his research work Schulz tried to prove that this arrangement was the basic rule of life. In 1888, for example, Schulz tested the effect of several poisons (sublimate, iodine, bromine, arsenic acid, chromium acid, salicylic acid and formic acid) on the cells of yeast with the help of an intricate experimental setting. He found that the greatest dilution of the substances caused an increase of yeast activity. Schulz now postulated that any stimulation of a living cell had an effect that was *inversely* proportional to the intensity of the stimulation.[136]

Homoeopathic thinking in medicine and pharmacology was supported by the success of homoeopathic pharmaceutical companies in the second half of the nineteenth century, and the Arndt-Schulz Law was often quoted around the turn of the twentieth century. To a certain extent, it bridged the gap between homoeopathy and academic medicine.[137] Despite criticism, the Arndt-Schulz Law survived and saw a new uptake after the First World War. This was a result of the rise of constitutional thinking and the 'crisis in medicine' in Germany in the 1920s.

Scientific medicine was criticized as materialistic and organ-centred, and the response was to increase support for holistic theories that analysed the health of people in their geographical and social environment.[138] The Arndt-Schulz Law fitted well into this general approach, and it is no surprise that it was supported by Friedrich Martius (1850–1923), professor of clinical medicine in Rostock since 1891, who was one of the pioneers of constitutional medicine in Germany.[139] In 1923, Martius published a paper in which he explained and defended Schulz's theories. Martius described Schulz as an unacknowledged genius and as a supporter of constitutional thinking, especially concerning the constitution of the individual.[140]

Martius also referred to another keen supporter of the Arndt-Schulz Law, the Berlin surgeon August Bier (1861–1949). Bier had held the chair in surgery at the University of Berlin since 1907 and indeed propagated the Arndt-Schulz Law. He had heard about homoeopathy when attending Schulz's lectures as a student, and later, at least from 1925, he promoted related ideas in Berlin. However, Bier's view that homoeopathy contained 'a good core idea' was severely attacked by his colleagues in the medical faculty in Berlin. Despite all the efforts of its supporters, the Arndt-Schulz Law was never generally accepted.[141]

Although the Arndt-Schulz Law corresponded to the fact that many drugs showed an effect in small doses and produced symptoms of intoxication in large ones,[142] it provoked criticism from pharmacologists. In 1923, Hans Handovsky (1888–1959), research assistant at the pharmacological institute of the University of Göttingen, collaborated on a paper about the effects of histamine on a parasite, the *protozoon balantophorus*. Histamine accelerated the growth of these one-cell animals in very small doses and killed them in larger ones. These findings in favour of the Arndt-Schulz Law soon turned out to be wrong. The growth acceleration was not due to histamine, but to a growth-stimulating substance released by dead protozoa.[143] In 1930, H. Dannenberg from the pharmacological institute in Berlin focused on experiments with yeast in a paper dealing with the question of the validity of the Arndt-Schulz Law. But he did not succeed in producing an accelerated fermentation with minimal doses of poisons such as quinine, phenol and sublimate and was unable to support Schulz's theories.[144] In 1933, A.J. Clark pointed out that the behaviour of the majority of drugs did not support the Arndt-Schulz Law and that evidence in favour of it could have been produced through experimental errors. Clark concluded that there was no serious reason to suppose 'that any such general law exists'.[145]

Conclusions

The existence of the competing theories of drug action discussed in this chapter contribute to an explanation of the fate of the receptor concept in pharmacology during the first three decades of the twentieth century. It shows that this concept was not easily accepted, although Langley as well as Ehrlich tried to promote their theories in diverse medical areas, including the community of pharmacologists. Two reasons can be suggested for this reluctant acceptance.

First, neither Langley nor Ehrlich could provide any direct evidence for their theories. The technical means for demonstrating receptors directly did not exist. This was a problem for all pharmacological theories that were constructed to explain the binding of drugs to organs or cells, and it was not soluble at the time. Any theory that ventured into this field remained, to a certain extent, speculative. The consequence was that every theorizer was vulnerable to the immediate appearance of serious critics. Ehrlich's receptor concept was especially provocative, because it was far-reaching and constituted a very general hypothesis about the metabolism of the human body, illustrated by his own drawings. Notwithstanding these problems, a widely perceived need to find explanations for the phenomenon of drug binding led to the positing of numerous different theories. As we have seen, they often had their roots in some nineteenth-century ideas or research strands, and they could be quite 'exotic', as the applications of the Weber-Fechner Law or the Arndt-Schulz Law to pharmacology show.

Second, the primary threat to the receptor concept, the physical theory of drug binding, was supported by powerful leaders of the emerging discipline of pharmacology, such as Walther Straub, Arthur Cushny and Walter Dixon. There is evidence that all three had considerable influence on their colleagues. And it was not by chance that they chose the physical theory of drug binding, which fitted well into the research tradition of the founding fathers of scientific pharmacology, Rudolf Buchheim and Oswald Schmiedeberg. Schmiedeberg especially had a large number of students, among them Straub and Cushny. In turn, Dixon was a student of Cushny. The Buchheim/Schmiedeberg School had propagated the experimental approach of scientific pharmacology along the lines of the methods of physiology. This also meant a special consideration of specific experimental settings and of physics in general, as new physical methods were the backbone of physiology in the nineteenth century.

Following the example of physiology implied the use of animal experiments, the aim of which was to investigate the physiological reaction of living beings to drugs in a very pragmatic sense. The question of how precisely the applied substances bound to organs or cells had been left open by Buchheim and Schmiedeberg. Their style of animal experimentation provided the guideline for the experiments of Straub, Cushny and Dixon. It was this style especially – the concentration on healthy animals and above all the neglect of the relationship of constitution, distribution and effect of a drug – that was criticized by Ehrlich. In contrast to the main representatives of pharmacology, the adherents of the receptor concept had a weaker methodological basis: while chemistry, the basis of the receptor concept, was practised in medical laboratories in the first half of the nineteenth century, it was only at the end of the century that it started to be applied to scientific medicine on a larger scale. Only around the turn of the twentieth century did what became known as clinical chemistry invade hospitals and become involved in the routine treatment of patients. Even after 1900, medical and clinical chemistry were still at a very early stage of implementation.[146]

It is not, therefore, surprising that Straub, Cushny, Dixon and others rejected Ehrlich's receptor theory as too far-reaching. It was easier for them to support the physical theory of drug binding because it was closer to the physically-oriented discipline of physiology that remained the guiding discipline for pharmacology in the twentieth century – above all in Britain. Moreover, with the acknowledgement of the fashionable field of physical chemistry, physical theorists in pharmacology could lay claim to new approaches in medicine.

4
Receptors and Scientific Pharmacology II: Critics of the Receptor Idea and Alternative Research Strands: the Transmitter Theory, *c.* 1905–35

As we saw in the last chapter, proponents of the chemical theory were confronted with the physical theory as a competing explanation of the specific mode of action of drugs. But they were also confronted with a competitive new research strand, namely the work on transmitter substances. Links existed between the receptor idea and the concept of transmitters within the notion of a 'receptive mechanism' (to use Sollmann and Hanzlik's term). However, this did not mean that they had the same history. On the contrary, the two research strands developed quite independently and the massive interest of pharmacologists in transmitters actually hampered the unfolding of research into receptors. In other words, the focus on transmitters decreased the probability that the hypothetical ideas about receptors and concepts of receptors would become a major issue in pharmacology. This is especially true for the first three decades of the twentieth century. In this chapter we take a closer look at this process.

First, we deal briefly with the history of neurotransmission. The next two sections introduce the ideas of two important physiologists/ pharmacologists performing transmitter research: Henry Hallett Dale (1875–1968) and Otto Loewi (1873–1961). This will provide the basis for discussing their attitude towards receptors in a fourth section. In the final section, we will analyse our findings and draw a conclusion.

The history of neurotransmission in the nineteenth century

Research on transmitters of the nervous system had been performed since the end of the nineteenth century and provided the basis of twentieth-century developments in this area. This research coincided with the micro-anatomical investigation of the nervous system. Primarily between 1889 and 1906 Santiago Ramón y Cajal investigated the structure and connections of nerve cells. Wilhelm von Waldeyer-Hartz introduced the term 'neuron' in 1891, and in 1890 Charles Scott Sherrington had examined the reflex responses of peripheral nerves and was able to identify the pathways of nerve impulses. Furthermore, he introduced the term 'synapse' for the point of contact between nerve ending and effector cell.[1] From the late nineteenth century, the autonomic nervous system was being investigated, particularly in Cambridge by the school of Michael Foster, and especially by John Newport Langley (see Chapter 2 above). In the winter of 1893–94 an important step was taken by the physician George Oliver and the London-based professor of physiology, Edward Albert Sharpey-Schäfer (1850–1935), who studied the effects of adrenal extracts (that is, extracts from the suprarenal glands) on animals and found constriction of the blood vessels and a marked increase of blood pressure. This research strand was followed up by others, including Langley, and the pattern of effects after administering the extract to animals was recognized to be similar to the effects following stimulation of the sympathetic part of the autonomic nervous system (see also Chapter 2). Eventually, after the turn of the century, the newly isolated substance 'adrenalin' (in the USA called 'epinephrine') was investigated. By 1915 it was clear that certain impulses were transmitted from one nerve to another, though it was not yet possible to elaborate on the specific way in which this worked.[2]

Henry Hallett Dale and the idea of transmitter substances

Such efforts to investigate substances related to the maintenance of the biochemical and physiological equilibrium of the human body were fashionable when Henry Hallett Dale came onto the scene in 1904. In the interwar years, Dale became one of the most influential physiologists and pharmacologists in Britain. He had received his medical education in Cambridge, starting in 1894, and in his student years he had come under the influence of Michael Foster and his school, especially Langley. From 1900 to 1902, Dale received his practical clinical training at St Bartholomew's Hospital in London. Having finished his medical education, he worked in London in the physiological laboratories of Bayliss and Ernest Henry Starling (1866–1929). Furthermore, Dale spent a few months with Ehrlich in Frankfurt. In 1904, Dale accepted – against the advice of friends – a post at the Wellcome Physiological Research

Laboratories, located in Herne Hill, South London. He thus made his career in a pharmaceutical company, becoming director of the laboratories in 1906. In 1914, Dale became a member of the scientific staff of the Medical Research Committee (after 1920 renamed the Medical Research Council), and between 1928 and 1942 he served as the director of the National Institute for Medical Research in London. Between 1942 and 1946, he was professor of chemistry and director of the Davy-Faraday Laboratories at the Royal Institution of Great Britain, and from 1942–47 he was chairman of the Scientific Advisory Committee to the War Cabinet. He received many honours, above all the Nobel Prize for Medicine or Physiology in 1936, which he shared with Otto Loewi, professor of pharmacology at the University of Graz.[3]

Dale's research – as well as that of Loewi – reflected the dominance of the transmitter concept in the first half of the twentieth century. It will be discussed below with specific reference to its relationship to the receptor concept. Dale's work is particularly interesting and informative in this respect, and we will therefore mainly concentrate on him. His work built upon contemporary investigations of the autonomic nervous system, and within this field it focused on the research into two substances: histamine and acetylcholine. The starting point was his research in the Wellcome Physiological Research Laboratories in 1904. His work there benefited from the generous financial support of Wellcome and from collaboration with outstanding colleagues, above all the chemist George Barger.[4]

At Sir Henry Wellcome's (1853–1936) suggestion, Dale examined a sample of ergot, and his research path developed from the investigations related to this fungus. Together with Barger, he worked on histamine from about 1907. They were able to isolate this substance as a component of ergot in 1911. Somewhat later, they found it as a common substance in the intestines of the ox, demonstrating that histamine was present in the animal body. In the years to come, Dale studied the role of this substance in allergic reactions; and in 1927, Dale and colleagues provided evidence that histamine could be found regularly as a messenger substance in animals. Only two years later, in 1929, he identified histamine as a substance that is responsible for anaphylactic reactions and shock. This led to further research on allergies and the defence reactions of the body.[5]

With regard to the question of receptors, Dale's work on acetylcholine was particularly important. The investigation of the ergot sample in 1904 marked the beginning of this research strand. It was known that ergot had an effect on the autonomic nervous system and Dale and Barger first thought that this effect of ergot was similar to that of adrenalin. Their views were supported by the research findings of Thomas Renton

Elliott (1877–1961), Dale's friend and a student of Langley, who in 1904 made the connection between adrenalin with the sympathetic part of the autonomic nervous system and its physiological effects: increase of blood pressure and of certain metabolic processes. Furthermore, Elliott had suggested that adrenalin was released by sympathetic nerve endings at their point of connection with muscle cells. He called this connection the 'myoneural junction' (see Chapter 2). Although Elliott did not follow up on this idea in the years to come,[6] experiments by Walter Dixon on the isolated frog's heart, in 1907, supported the idea: when the vagus nerve was stimulated, a substance would be set free and would combine with a component of the cardiac muscle, thus causing an inhibitory effect.[7]

However, the difficulty for Dale was to reproduce the 'adrenalin-like' effect of ergot routinely in animal experiments. In 1913, by chance, he detected another component of ergot: acetylcholine. The routine testing of one sample had caused the heartbeat to stop and the collapse of circulation in a cat. So the component had the inverse effect of adrenalin and corresponded to the effects of the parasympathetic part of the autonomic nervous system. One of Dale's chemists, Arthur James Ewins (1882–1957), was able to isolate the substance, acetylcholine. While the vasopressor effects of acetylcholine had been examined by Reid Hunt (1870–1948) in 1906, it had not been identified in animal bodies and had until then been known only as an artificially produced substance. Now there was the suspicion that acetylcholine was a regular messenger substance in animal and human bodies but there was no evidence for this hypothesis, and the First World War led to an interruption of efforts to clarify the point.[8]

Otto Loewi and neurotransmission

Further input came from Dale's friend Otto Loewi, who worked on problems of human metabolism, the functions of the autonomic nervous system and the transmission of nervous impulses. Loewi had studied medicine in Strasbourg and Munich between 1891 and 1896. After this he undertook further training in inorganic analytical chemistry and in clinical practice. In 1898, he became assistant to Hans Horst Meyer at the University of Marburg. During his time in Marburg, he visited London and Cambridge, working with Starling and Langley, and he made a lifelong friendship with Starling and Dale. In 1905 he spent some research time at the Zoological Research Institute in Naples, and in the same year he moved with his teacher Meyer to Vienna. In 1909 he became professor of pharmacology at the University of Graz. Here, Loewi did his most important scientific work, although under unfavourable conditions since the building in which the pharmacological institute was housed had no modern technical facilities and inadequate scientific equipment. But Loewi remained in Graz until 1938, when he left for London following the seizure of power by the National Socialists. In 1940 he became a professor in New York, where he stayed until his death in 1961. Like Dale, Loewi was an honoured scientist (receiving the Nobel Prize together with Dale in 1936) and he had had prominent teachers: Schmiedeberg in Strasbourg and Meyer in Marburg and Vienna. This bound him in some way to the nineteenth-century approaches of German experimental pharmacology. In Strasbourg he met Arthur Cushny, and it was here that a lifelong friendship with Walther Straub began.[9]

The exchange of ideas with English colleagues, above all with Elliott, led Loewi to investigations into the effects of foreign and bodily substances on the nervous system. Between 1913 and 1921 he worked on the influence of calcium and digitalis on the isolated frog heart. Initially without success, he tried to find an explanation for the mechanisms of the transmission of nerve impulses. His investigations of 1921 were a decisive step forward in this direction, and also influenced the post-war work of Dale. Loewi performed a simple experiment, based on an equally simple experimental setting: the starting point was two frog hearts, which were kept in a fluid. Loewi stimulated one of these isolated frog hearts, subsequently transferring the fluid of this heart to the second one, whose vagus nerve had been removed. Depending on the stimulation of either the activating or the depressing part of the vagus nerve of the first heart, the second heart showed the same effects, that is, either an increase or a decrease of the heart rate. This was evidence that a substance, which he simply called 'vagus substance' ('*Vagusstoff*'), served as a messenger for the nerve's impulses; he named its antagonist 'accelerance substance' ('*Acceleransstoff*'). The neural transmission was thus of a chemical nature. In 1926, Loewi and his colleague, Navratil, suggested that their 'vagus substance' was acetylcholine.[10]

Encouraged by Loewi's results, Dale intensified his own research on acetylcholine and in 1929, together with a colleague, he successfully isolated this substance from the spleen of the ox and the horse. In the years to come, Dale found acetylcholine in different parts of the autonomic nervous system and in 1933 he introduced the terms 'cholinergic' and 'adrenergic' synapses, depending on transmission through acetylcholine or adrenalin.[11]

Dale, Loewi and the receptor concept

Even this short account of the pioneering research on acetylcholine and of the main aspects of the work of Dale and Loewi provides some indication of their attitude towards the receptor concept, especially during the first decades of the twentieth century. While both scientists were directly in contact with Ehrlich and even more so with Langley, Ehrlich's 'receptors' and Langley's 'receptive substances' were not Dale and Loewi's primary focus and the receptor idea remained a marginal topic in their publications. They were mainly interested in the other part of the mechanism of nervous impulse transmission, namely transmitter substances, and they selected this research field from all those research areas that they had been introduced to in Frankfurt, Cambridge and London. Dale, for example, in his Nobel lecture, saw his early research on the anatomy of the autonomic nervous system in Cambridge as the only influence that his teacher Langley had on his work.[12]

In spite of the overlap in their research interests, Dale and Loewi dealt with the question of receptors in different ways. In an age of growing demands on investigations into the complex biochemistry of the human body, Dale considered the receptor idea as a hypothesis which had seriously to be taken into account. However, its speculative nature remained problematic for him. Generally, Dale was a pragmatic researcher rather than someone who constructed intricate theories: 'Dale was by temperament reluctant to speculate much beyond what the evidence in hand could firmly support' and tended rather 'to apply the brake than to be the first in the gold rush'.[13] In the tradition of nineteenth-century pharmacology he was chiefly interested in the observable effects of substances on organs and organisms, following a two-step procedure when applying a substance to a research animal and then investigating the symptoms. Two examples may serve to illustrate this. In 1905, he investigated the effect of nerve stimulation and of the application of various substances on the gall-bladder of dogs. In 1926, he applied acetylcholine and other substances to denervated muscles of the cat to examine the effects. This style of experimentation reflected his commitment to the physiological research that formed the backbone of early twentieth-century pharmacology. Dale argued against bringing all pharmacological approaches under one theoretical scheme and preferred to leave gaps of knowledge until there was more evidence. In his view, animal experimentation on the question of transmitter substances was a solid research field that promised therapeutic applications. These applications were more doubtful in the case of the receptor theory.[14]

Against this background, Langley's theory of 'receptive substances' was one of the major targets of Dale's criticism. Dale's antipathy towards the Cambridge physiologist was to some extent the result of his opinion that Langley theorized too much.[15] In 1910, Dale and Barger examined the chemical structure and sympathomimetic action of amines. They were not successful in correlating the chemical structure of the amines with their specific effects and were critical of the chemical theory of drug action. It followed by no means

that the peculiar distribution of the action of nicotine or of the sympathomimetic amines depends on the existence of specific chemical receptors in the cells peculiarly sensitive to them, as supposed by Langley. Stimulation may be a chemical process: but the fact that certain cells are preferentially stimulated by a certain group of substances, such as our amines, may mean that in those cells these substances easily reach the site of action; a supposition which is in accordance with the view advanced by Straub.[16]

Although their paper argued in remarkable detail in favour of the physical theory, and against Langley, Dale's problems with theorizing were apparent in his concluding remarks: 'Neither the purely chemical nor the mainly physical view satisfactorily accounts for the production of inhibition and motor action in different tissues by the same, or in the same tissue by different bases.'[17] In a letter to Elliott, written in 1913, Dale again articulated his problems with far-reaching theories, and he was also much more explicit in his criticism of Langley: 'Langley's attitude on the whole question seems to me simply silly. His own ideas are absolutely muddled, and I cannot avoid the impression that he deliberately uses vague and meaningless expressions, in the hope that the general crowd will be awed by the sense of mysteries beyond their intelligence.'[18]

Dale's hostility to theorizing in general, and to Langley's theory of drug action in particular, had already become apparent in 1906, when he had given credit to Elliott for results achieved in the examination of ergot, seeing the latter's theory of the myoneural junction as sufficient to explain specific effects of substances on animal organs. He used Elliott's terminology, and with clear reluctance to accept Langley's theory, Dale noted that if 'Langley's recent terminology be accepted' one would have to use the phrase 'receptive substance for adrenalin' instead of 'sympathetic myoneural junction'. Remarkably, in the end Dale left the question open as to whether there were 'myoneural junctions' or 'receptive substances'.[19]

This episode indicates that in the long run, Dale did not exclude his friend Elliott from his criticism. Elliott's 'myoneural junctions' seemed closer to Langley's receptive substances than to the presynaptic nerve ending which was the focus of Dale's interest. After all, Elliott believed in the existence of a sensitive and responsive region of the muscle cell.[20] In Dale and Barger's 1910 paper, Elliott was criticized in particular because of his hypothesis that the nerve endings might liberate adrenalin.[21] And in 1935, Dale summarized both Elliott's and Langley's concepts under the category of doubtful theories, although Elliott himself was no supporter of Langley's receptor theory. Both Elliott's 'myoneural junctions' and Langley's 'receptive substances' were, in Dale's opinion, 'hypothetical components of the effector cells' which no longer served 'to clarify the issue'. In Elliott's case the term implied 'a localization of the specific excitability' in the neighbourhood of the nerve endings of involuntary muscle cells, for which there was no evidence. In Langley's case there was no evidence for 'a chemical fixation of the stimulating substances'. Furthermore, Dale reiterated his argument that there was no close relationship between chemical structure and the pharmacological action of a substance.[22]

Dale also targeted Ehrlich. In 1910, he and Barger remarked that the 'theory of receptive side-chains is, indeed, very difficult to apply to our results with such precision as ought to be possible in the case of simple substances of known constitution'.[23] In 1920, he criticized 'that wonderful fabric of theory with which Ehrlich so largely influenced the form and direction of research in pathology and pharmacology for almost a generation'. The challenge for Dale was 'to replace, by a rational conception, the pictorial schema of its [the antibody's] relation to the antigen, which Ehrlich displayed in the side-chain theory'. Dale confronted Ehrlich's theory with arguments from physical theories, such as the 'complex colloidal system' or – with explicit reference to Cushny – the impact of optical activity. But as there would not be 'one principle of interpretation', Dale argued against relying on 'stimulating suggestions and seductive analogies' and for achieving more knowledge under 'the guiding influence ... from the physiological side'.[24]

Dealing with recent trends in chemotherapy, Dale criticized Ehrlich again only a year later, in 1921. His side-chain theory was no longer acceptable 'as a scientific presentation'. His theory needed to be 'modified' with the help of the new insights from colloidal chemistry. Dale's most important argument against the side-chain theory was that the interaction between the microorganism and the host had been neglected, and he pleaded for empirical clinical observation of patients.[25] This last argument was repeated two years later, in 1923, when Dale again made critical remarks about the side-chain theory, not only arguing that there was no isolated relation between parasite and drug, but also that the host influenced the healing process in an infectious disease. Furthermore, Dale's pragmatic attitude towards experimental problem solving collided with Ehrlich's speculative microcosm of receptors (see Chapter 1). Dale accepted the idea that different organisms of the same species or of related species had different receptors for different substances, although 'it does not convey more real information than the record of the results observed', but the idea that different parasites had the same receptors for certain drugs was simply absurd for him. Things were far more complicated in his view, and he argued for an integration of chemotherapy into 'a new theoretical foundation' of drug therapy.[26]

Dale did not only criticize the main representatives of the receptor concept, but expressed his own visions of the goals of modern pharmacology in several papers published between 1930 and his Nobel lecture in 1936. Remarkably, the receptors play almost no role in Dale's conceptions. He emphasized the collaboration between physiology and clinical medicine and the shift from symptomatic treatment to the specific application of therapeutic substances. This said, he mainly praised the achievements of biochemistry, especially regarding the functions of the endocrine glands and the detection of vitamins and hormones. Dale mentioned the treatment of diabetes with insulin, which was seen in his time as a revolutionary therapeutic step, and the treatment of thyroid disorders with thyroid extracts. For him, therapeutic progress was closely linked to the investigation of messenger and transmitter substances. These were the fields to which he and Loewi could contribute with their research on neurotransmission. This research consisted chiefly in investigation of transmitting substances. 'Receptive cells' were rarely mentioned, and only insofar as the effects of the transmitter substances were concerned. The 'receptiveness' as such was not interesting and was left aside as a black box.[27] This is important also insofar as Dale in these days was a trendsetter for new research directions in pharmacology. His mostly critical comments promoted the tendency to push the receptor concept to a minor, side branch of research in pharmacology and presented an obstacle to its recognition. This means that the historical interpretation that the development of beta-receptor blocking drugs[28] 'can be seen as an

almost direct continuation of Dale's very early observations on the reversal of the effect of adrenalin by an extract of ergot' cannot be supported
without considerable qualification.[29]

Otto Loewi reacted very differently to the challenge of the receptor
concept. Of course, like Dale, from early on Loewi was aware of the different theories explaining the attachment of bodily and foreign substances
to cells. But in contrast to Dale, Loewi needed such theories, because in
his view it was evident that some kind of binding of these substances
to organic structures happened in the body. Loewi strongly believed in
chemistry as a key to the understanding of the physiology and pathophysiology of the human body. He used both the chemical theory and
the physical theory to explain substance binding to cells. In 1902, in a
paper on the synthesis of proteins in the animal body, Loewi stated that
certain components would split off from proteins and 'bind' to the wall
of the bowel. Furthermore, certain 'binding bodies' (*'Bindekörper'*) would
circulate in the bloodstream and release protein components, depending
on the requirements of the organs. Although there was no certain knowledge in this area, Loewi reminded the reader that Ehrlich, on the basis of
his experiments on haemolysis, had made comments on the metabolism
of nutritive substances (*'das Wesen der Nahrungsassimilation'*).[30]

In the same year, Loewi published a paper on the physiology and
pharmacology of kidney function in which he re-emphasized that there
was only meagre understanding of the 'physical and chemical components' (*'physikalische(n) und chemische(n) Componenten'*) of the 'specific
actions of the cell' (*'specifische Zellthätigkeit'*). However, physical chemistry served the purpose of providing a provisional explanation of binding processes: waste products would circulate in the blood in 'colloidal
binding' (binding of particles in solvents via adsorption and surface
tension).[31] This set the stage for Loewi's maintenance of an ambivalent
attitude towards the theoretical basis of substance binding to cells, at
least until the end of his productive period as a researcher, that is, until
he left Austria in 1938. In 1910, he wrote that the effect of pilocarpine
and atropine on the isolated muscle of the toad depended on the 'mixture of ions' and influenced the 'permeability' of the cellular membrane,
thus using the vocabulary of the physical theorists.[32] In a paper on his
famous experiments on the '*Vagusstoff*', he stated that there were 'chemically or physically characterized parts... of the end organ' (*'chemisch
oder physikalisch charakterisierte Teile ... des Erfolgsorganes'*), which he now
labelled, using Langley's term, 'receptive substances'. The substances of
the vagus and the sympathicus (*'Vagus- und Sympathicusstoffe'*) exerted
their effects not on the nerve but on these parts of the end organ.[33]

In 1926, in the tenth of his series of papers on the '*Vagusstoff*', Loewi wrote about 'physical-chemical changes' of the extract ('*physikalisch-chemische Änderung des Extraktes*') and again applied a term used by the physical theorists and colloidal chemists when mentioning changes of the 'condition of adsorption' ('*Adsorptionsverhältnisse*'). He used the term 'adsorption' also in other papers, in 1927 and 1930.[34]

Loewi's consideration of chemical as well as physical (or physico-chemical) theories of drug binding did not mean, however, that he had any great interest in such theories. In contrast to Dale, to whom these theories (especially the chemical theory) appeared to constitute a kind of offence or threat to solid pragmatic research, Loewi commented on them only rather briefly and vaguely. Essentially, he left open the question of which theory of drug action was more appropriate, specifically mentioning the competing theories in one or two sentences, or at the most in a very short paragraph. Typical is Loewi's remark in a paper of 1924 on the site of action of atropine. The substances released by nerves into the fluid of the heart would 'attack' ('*angreifen*') 'somewhere at the end organ itself' ('*irgendwo am Erfolgsorgan selbst*'). The question of the way in which the substances actually bound to the cells was not discussed.[35] Likewise, in the same year, Loewi and a colleague wrote about the effect of iodine on the respiration of isolated cells without exploring the question of substance binding.[36]

In the second half of the 1920s, Loewi published several papers on the effect of insulin and the metabolism of glucose in the human body. In this context, he discussed the fixation of glucose ('*Glucosefixierung*') to plasma and erythrocytes and in general the attachment of glucose to organic structures ('*Strukturfixation*'), yet without making any comments about the details of how glucose was attached to other substances or cells. It was self-evident 'that ferment and substrate have to come together' ('*daß Ferment und Substrat zusammenkommen*').[37] As far as neurotransmission was concerned, the question of the way in which the '*Vagusstoff*' attached to the end organ was not dealt with. Even in his Nobel lecture of 1936 (which he used to summarize his work), Loewi avoided the topic.[38] The word 'receptor' was used only in connection with certain cell organs which were able to receive certain stimuli, such as the tactile receptors of the skin. Loewi published one short article on these kinds of 'receptors', namely on those of the digestive channel.[39]

The reason for Loewi's reluctance to deal with 'receptors' or 'receptive substances' in Ehrlich and Langley's sense probably lay in his conviction that 'neural transmission' at the last stage was simply the release of chemical transmitters by nerve endings and their effect on the end organ. Loewi focused his investigations on transmitter substances in the autonomic nervous system, and he gave precedence to quoting Langley in connection with the latter's work on the effects of those substances.[40] It was from this perspective that Loewi interpreted research data. If, for example, the degenerated sympathetic nerves of the pupil of the cat's eye widened after the application of adrenalin, he did not consider receptive substances in the end organ. By contrast, he believed that a certain inhibiting function of the nerve – unknown so far – had been lost.[41] And in his Nobel lecture he described the synapse as only important for nervous impulse transmission.[42] Loewi's attitude towards the question of receptors is summarized very well in a comment on the effect of transmitters on the end organ, which he made in an article in 1937:

> As with the other alkaloids, we do not know the definite mechanism of the transmitters ... The site of attack of substances is after all some chemical or chemical-physical component of the end organ, namely of the cell itself. The substances have their effect on the cell, as far as we know on the surface of the cell and only from outside ... That the effect of a particular substance mainly occurs at the site of its release, that the cells are sensible for the substance especially at this site, is a phenomenon which is part of the specific sensibility of the living organism towards well defined chemical substances that are necessary to survive, a sensibility which is not understandable causally but only in a teleological manner.[43]

By 1937, four years after A.J. Clark's publication of a new, quantitative approach to receptors, and 11 years after the publication of Sollmann and Hanzlik's receptor concept, Loewi still held to the view that nothing could be said about receptors, that the phenomenon of the attachment of transmitters to the end organ might be chemical or chemical-physical, and that the transmitters should be the focus of interest for research into nervous transmission. Like Dale's attitude, this view constituted an obstacle to further work on the receptor concept.[44]

Conclusions

The majority of the scientific community of pharmacologists competed with the supporters of the receptor concept not only on the basis of the alternative concept of the physical theory of drug action but also with an alternative research strand, namely the transmitter concept.

The difficult problem of how drugs bind to organs or cells could be circumvented when concentrating research on the transmitter substances themselves as the decisive agents for the mediation of drug effects on the animal and human organism. Measuring the vegetative reactions of the animal body to the application of transmitter substances also corresponded to the Buchheim/Schmiedeberg tradition. The conceptual overlaps between physical theorists and transmitter researchers are obvious: Henry Hallett Dale saw himself as a physiologist and, together with George Barger, he favoured the physical theory of drug action at least at the beginning of his career. On this basis, transmitter research allowed for the development and evaluation of therapeutic drugs. This met the demands of early twentieth-century medicine, enabling the application of knowledge acquired in the preceding century to medical practice.

On the basis of Dale's and Loewi's research it was possible to develop drugs in the interwar period to combat unwanted vegetative symptoms and transmitter related diseases. After 1934, the disease myasthenia gravis, a muscular weakness caused by a shortage of acetylcholine, could be treated with the alkaloid physostigmine and then with neostigmine. These drugs suppress the release of acetylcholine esterase – an enzyme that destroys acetylcholine. Concerning their usage in therapy it was sufficient to know that physostigmine and neostigmine influenced the metabolism at the myoneural endplate.[45] In contrast to clinically-oriented transmitter research, basic research on potential receptors seemed an unpromising field. Neither Dale nor Loewi supported the receptor concept. Dale polemicized against Ehrlich, Langley and Elliott, and his position as one of the leading scientists in pharmacology meant that those who promulgated the ideas of receptors or a receptor concept had severe problems establishing their approaches within the scientific community. Only after 1945, as a result of changes in the scientific environment and other developments (see the following chapters), was the receptor concept gradually accepted as an explanatory model for drug binding. Present-day thinking is that myasthenia gravis is caused by a shortage of acetylcholine receptors due to autoimmune antibodies.[46]

Against the background of the last two chapters and in the context of the reaction of scientific pharmacology to the creation and propagation of the receptor concept, we can make one further point that helps our understanding of the problems that receptor theorists faced. The scientific community of pharmacologists, which favoured the physical theory of drug binding and/or the transmitter concept, built a network which made it hard or, indeed, impossible for outsiders who supported the receptor concept to influence the direction of research. Cushny and Straub knew each other well and were friends. The same is true for Straub and Loewi, who was a friend of Dale. And Straub also had good relations with William Maddock Bayliss, who was a teacher of Dale's.[47] Cushny was a close friend of Hans Horst Meyer, who taught Loewi, and Cushny had worked with Loewi in Schmiedeberg's laboratories in Strasbourg. This considered, it is not surprising that there was a certain reluctance to accept new scientific approaches that might cause major shifts in the research principles and practices of the discipline.

Actually, it is remarkable that the receptor concept was able to survive, but survive it did. Certainly, one reason for this was the inability of the physical theory and of the transmitter concept to explain all steps from the application of a substance to its binding to organic substrate. The physical theory was criticized as being applicable only in specific experimental settings, and its results were far from sufficient to construct a general law of drug binding to cells; and the transmitter concept could describe and analyse the character of active substances, but not how these substances react with the substrate of the cells. Dale and Loewi left this question open and postponed further research on it. These weaknesses of the physical theory and of the transmitter concept enabled the chemical theory gradually to gain a foothold even within the work of the most prominent pharmacologists; Straub, Cushny and Dixon were all eventually forced to give chemical explanations more and more credence. And for Loewi, chemistry was a key to solving research problems on the metabolism of the human body, although he did not apply it to the receptor question. In 1928, Torald Hermann Sollmann and Paul John Hanzlik even developed a theory on the 'receptive mechanism' – the first new receptor concept since those of Langley and Ehrlich. The situation was very open to the development of more detailed drug binding theories and around the middle of the twentieth century elaborated receptor concepts were constructed by Alfred Joseph Clark (1933) and Raymond P. Ahlquist (1948).

5
Quantitative Arguments for the Existence of Drug Receptors and the Development of the Receptor Occupancy Theory, *c.* 1910–60

On the basis of animal experiments with antagonistic drugs and with suprarenal extract ('adrenalin'), in 1905 J. N. Langley proposed the existence of specific 'receptive substances' in cells. When drugs or hormones bound chemically to these substances, the metabolism of the 'chief substance' of the cell would be influenced, and thus the cell's function (for example, secretion in gland cells, contraction in muscle cells) would be changed. In this manner, Langley suggested, adrenalin contracted the smooth muscle of blood vessels, nicotine led to contraction of skeletal muscles and curare to their relaxation, pilocarpine stimulated secretion of saliva and atropine stopped this secretion (see Chapter 2 above). Paul Ehrlich, who had developed his 'side-chain' or 'receptor' theory in the study of toxin binding to cells and antitoxin (antibody) production a few years earlier, soon adopted Langley's view, introducing, in 1907, the term 'chemoreceptors' for those specific constituents of the cell that were able to bind certain chemical substances. Ehrlich's chemotherapeutic research on trypanosomiasis (sleeping sickness) and syphilis rested on the assumption that certain drugs, such as dyes and arsenical compounds, were fixed selectively to the pathogenic microorganisms (leading to their death), while they left the host's body cells largely unharmed (see Chapter 1 above).

These early ideas about drug receptors had essentially been of a qualitative nature. However, the relationship between the dose of a drug and its biological effect had also to be taken into account in order to use a drug safely and effectively in therapy. As the Edinburgh pharmacologist, Alfred Joseph Clark (1885–1941), observed in 1933:

> Pharmacology is one of the youngest of the biological sciences, and its youth was naturally occupied by the task of determining accurately

the qualitative nature of the action of drugs. During the present century more and more attention has been paid to the quantitative aspect of pharmacology. Such studies have been stimulated in particular by the remarkable advances made in chemotherapy and also by the necessity for finding methods for estimating the activity of drugs by biological tests, and in consequence the general problems of quantitative pharmacology are beginning to receive more attention.[1]

In this chapter we trace some major conceptual changes in drug receptor theory that arose from *quantitative* approaches to the study of drug action. Besides Clark, several other leading pharmacologists of the twentieth century, including John Henry Gaddum (1900–65), William D.M. Paton (1917–93) and E.J. Ariëns, contributed to this research strand.[2] Yet we also consider the continuing criticism and scepticism vis-à-vis the receptor idea. By the early 1960s, quantitative investigations of drug action and interpretations of the experimental findings in terms of the receptor concept had become central to the emerging field of pharmacodynamics. Even then, however, receptors were still merely hypothetical entities.

Early quantitative work on drug–receptor interaction

A first attempt to provide a quantitative basis for Langley's concept of 'receptive substances' was made as early as 1909 by one of his students, Archibald Vivian Hill (1886–1977).[3] This attempt was stimulated, in part, by the need to address competing *physical* theories of drug action (see Chapter 3). At Langley's suggestion, Hill studied the mode of action of nicotine and curare on the isolated *rectus abdominis* muscle of the frog. His mathematical analysis of the size and time course of contraction curves for nicotine and of relaxation curves for curare supported the hypothesis that these curves reflected 'a gradual combination of the drug with some constituent of the muscle', and not, as physical theories of drug action would have suggested, 'a gradual diffusion of the drug in or out' of the muscle preparation. Moreover, the temperature coefficient of the velocity of the nicotine contraction indicated to Hill that the drug's combination with the 'combining constituent' or 'receptive substance' of the muscle cells was 'of an ordinary chemical nature'. Accordingly, he concluded that the action of the drug on the muscle could be understood as a reversible chemical combination of the nicotine molecule N and the 'constituent' A of the muscle. The effect – in this case the intensity of contraction – would then be proportional to the amount of compound

(*NA*) formed between the two, minus a minimum or threshold amount (*M*) of compound that was necessary to produce any contraction at all. Curare would likewise combine chemically with a constituent of the muscle cell, producing relaxation instead of contraction.[4]

With this theory Hill especially opposed the physical theory of drug action that had been proposed by Walther Straub (see Chapter 3). According to Straub, the biological effect of a drug depended much less on its chemical constitution and structure than on its physical properties, which determined how it was distributed in the body and to what extent it was able to 'dissolve in' and to penetrate the outer membrane of cells. For him, the physical deformation of the cell membrane was the basis of pharmacological action. From experiments with muscarine on the isolated heart of the sea snail, the ray and the frog, Straub had proposed his 'poison-potential theory'. According to this theory, a drug or poison acted only as long as it diffused across the cell membrane due to a 'concentration potential' between the outside and the inside of the cell.[5] Langley, by contrast, had suggested that drugs reacted *chemically* with specific constituents of the cell, the 'receptive substances', to initiate a biological effect.

Hill's evidence for a reversible, monomolecular chemical reaction between drug and receptor was not followed up until the 1920s. The receptor concept itself was controversial and continued to be in direct competition with Straub's physical 'poison-potential theory' as well as other theories of drug action (see Chapter 3). Also, Hill's mathematical approach to physiological problems seems to have been perceived as rather unusual. While some colleagues in the Cambridge Physiological Laboratory made use of his mathematical abilities to provide theoretical foundations for their own experimental work, Hill recalled that his fellow physiologists had also joked about the fact that he had taken the mathematical tripos, 'just as though I had once taken a degree in theology or Sanskrit'.[6] In one of his next research projects Hill, together with his senior colleague Joseph Barcroft (1872–1947), similarly characterized the formation of oxyhaemoglobin as a monomolecular, chemical union – between haemoglobin and oxygen. He also studied oxygenbinding when haemoglobin molecules aggregated. In this context he provided the first kinetic model for the 'cooperative reaction' between several agonists (oxygen molecules) and a macro-molecular receptor (aggregated haemoglobin).[7] Hill did not, however, directly pursue the pharmacological issue of specific drug receptors. After 1910 he began to focus his research on muscle physiology, which in 1923 earned him, together with Otto Meyerhof (1884–1951), the Nobel Prize for Medicine or Physiology.[8] As for the general state of knowledge about specific drug action, Hill expressed a critical view in 1925, in a lecture as professor of physiology at University College London:

> In some ways the chemical structure of the living cell is analogous to the geometrical structure of a lock of almost infinite complexity. Substances similar in all their physical, and in nearly all their chemical, relations may show entirely different properties in their behaviour with living cells; and although the study of the action of drugs is one of the oldest of all the branches of science, we still have practically no clue as to the manner in which drugs exert their amazingly specific properties ... We need to know the actual chemical structure, the geometrical structure, of the substances which make up this membrane [of the cell], be they lipoid substances, or proteins, or something even more complex. We need to know the organic chemistry, the structural chemistry of the drugs, we require a geometrical picture to enable us to see how the lock is fitted by the key.[9]

It was not Hill, but another of Langley's students, Alfred Joseph Clark, who then contributed to receptor theory. As part of his medical education Clark had studied the natural science tripos at Cambridge between 1903 and 1907, coming thus under the influence of Langley and colleagues in the Physiological Laboratory. He had particularly good relations with Barcroft, who was then demonstrator in physiology. However, he was especially attracted to the field of experimental drug research through the work of Walter Ernest Dixon.[10] Clark received his clinical training in St Bartholomew's Hospital in London, and took the MB and ChB examinations in 1909 at Cambridge. After medical practice as a house surgeon at Addenbrooke's Hospital in Cambridge and as a general physician at St Bartholomew's, in 1911 he received a two-year research scholarship from the British Medical Association at King's College, London, where Dixon had been appointed to a professorship. Inspired by Langley's 'receptive substances' and Ehrlich's 'chemoreceptors', in 1909 Dixon had examined the receptor hypothesis for the case of the specific action of strychnine on the spinal cord. While his findings in these experiments did not support the notion of specific receptors for plant alkaloids such as strychnine, another line of Dixon's research, on the duodenal hormone secretin, suggested the existence of receptor substances for hormones (see Chapters 2 and 3).[11]

Clark spent the second year of his scholarship with Arthur R. Cushny, then professor of pharmacology at University College London (see

Chapter 3). Cushny had also taken up Langley's concept of 'receptive substances' and applied it in his interpretation of his experimental findings on the antagonistic action of pilocarpine and atropine on the uterus.[12] Generally, however, Cushny leaned more towards a physical theory of drug action. While he accepted that drugs attached to receptive substances in cells, he held that their pharmacological effect was chiefly determined by their physical characteristics. For example, he explained the differences in pharmacological activity between pairs of optical isomers of a substance through different physical properties of the two kinds of isomer-receptor complexes formed.[13] Unsurprisingly, therefore, Clark was interested in the question of drug binding to cell constituents from early on. However, he was also influenced by Straub's theory of physical drug effects on the cell membrane, and thus initially open-minded about the rival models of pharmacological action.[14]

In 1913, Clark was appointed to a lectureship in pharmacology at Guy's Hospital, London. In the following year he was awarded a Cambridge MD, based on a thesis about the detection of haemoglobin in faeces, a topic which he had studied during his time at Addenbrooke's. His early pharmacological work was concerned with the effects of digitalis glycosides and other substances on the isolated frog heart, a preparation on which he became an expert.[15] Clark's career in pharmacology was interrupted by the First World War, in which he served as an officer in the Royal Army Medical Corps. In 1918, however, he was appointed professor of pharmacology in the new medical school of Capetown University, and in 1920 he succeeded Cushny (who had moved on to Edinburgh) as the chair of pharmacology at University College London.[16] Clark's major contribution to the receptor theory was linked with his research on acetylcholine, which he carried out at UCL.

In the mid-1920s, acetylcholine was already suspected to be a transmitter substance in the nervous system. In particular, it appeared to be the '*Vagusstoff*' that had been characterized by Otto Loewi in 1921 as a substance that was produced upon stimulation of the vagus nerve and slowed down the action of the heart.[17] Clark regarded acetylcholine as especially suitable for the study of drug action on cells, because it produced reversible and graded effects over a wide range of concentrations, and because its effects differed with different tissues. He experimented with acetylcholine on isolated ventricular strips from the frog heart, in which it diminished the force of contractions, and on the *rectus abdominis* muscle of the frog, where it increased contraction.[18]

From both preparations Clark obtained data for the effect of acetylcholine that were lying along a sigmoid shaped curve when the action (in per cent of maximum action) was plotted against the logarithm of the concentration of the drug. Such a curve could be fitted closely with the equation $Kx = y/100 - y$, in which x was the molecular concentration of acetylcholine, K a constant, and y the pharmacological action produced, expressed as percentage of maximal action. Moreover, the amounts of acetylcholine necessary to produce an effect on the frog's ventricular strip were very small, and there was, contrary to Straub's 'poison-potential theory', no direct relation between the amount of acetylcholine *entering* the heart muscle cells and the action produced. Clark calculated the minimal amount of acetylcholine uniting with a heart muscle cell that would just be sufficient to produce a demonstrable action. He arrived at about 20,000 acetylcholine molecules per cell – an amount that was not enough to form a continuous layer over the heart cells or to cover a larger area inside them. The drug molecules would therefore be fixed only to a 'very small fraction' of the cell's surface. Drawing these observations together, Clark suggested that the simplest way to explain the concentration-action formula $Kx = y/100 - y$ was to assume that a reversible, monomolecular reaction occurred between the drug and some receptor in the cell or on the cell's surface.[19]

This was in essence the same conclusion that Hill had reached more than 15 years earlier from his study on the action of nicotine and curare on striated muscle. Unsurprisingly, Clark also pointed out the similarity in his action curves for acetylcholine and the dissociation curve for oxyhaemoglobin.[20] In fact, it is possible to construct an intellectual 'genealogy' of drug receptor theorists from Langley to Hill to Clark.

Ideas about drug antagonism and receptor occupancy

One issue on which the receptor theory of pharmacological action shed new light was that of the mechanism underlying the antagonism between certain drugs. The scientific debate over the nature of drug antagonisms reached back to the nineteenth century (see Chapter 2). In the early twentieth century, Straub provided an interpretation of drug antagonism on the basis of his 'poison-potential theory'. The antagonistic effects on the heart produced by muscarine (inhibition) and atropine (stimulation) could, in his view, be explained by the hypothesis that atropine slowed down the absorption of muscarine through the cell membrane into the interior of the heart cells. Langley, by contrast, defending his receptor concept against Straub, suggested that atropine combined with the 'receptive substances' of the heart cells and in this way inhibited the effect of muscarine.[21]

This controversial matter was further investigated by Cushny, who examined the antagonism between atropine and pilocarpine in the salivary secretion of the dog. In these experiments he obtained a constant ratio between the concentrations of the two antagonistic drugs that together were necessary for producing a defined effect. On this basis, Cushny suggested that the atropine occupying the 'receptors' of the salivary gland cells was 'displaced' by its antagonist pilocarpine. With reference to Ehrlich's work on toxin-antitoxin interaction, he further proposed that: 'According to current views the antitoxin withdraws the toxin by forming a combination of toxin-antitoxin, leaving the cell free, while pilocarpine may be supposed to oust the atropine from the cell and combine in some way with the latter, leaving the atropine free.'[22]

However, in parallel to this interpretation, Cushny also considered Straub's theory of drug antagonism and left room for the possibility that atropine might also reduce the permeability of the cell membrane for pilocarpine.

In the 1920s, John Henry Gaddum also produced new experimental evidence supporting the receptor model of drug antagonism. Like Clark, Gaddum had studied natural sciences under Langley in Cambridge. After his clinical training at University College Hospital, London, he qualified MRCS, LRCP in 1924. In the following year he joined John William Trevan (1887–1956) as an assistant at the Wellcome Physiological Research Laboratories, and in 1927 he became assistant to Henry Hallett Dale at the National Institute for Medical Research.[23] In the mid-1920s Gaddum examined the action of adrenalin on the isolated uterus of the rabbit, arriving at a sigmoid curve if he plotted the intensity of contraction against the logarithm of the drug concentration. As he saw it, this curve represented the action of adrenalin on 'a number of units [of the muscle] whose susceptibility is distributed about a mean in accordance with a probability curve'.[24]

In other words, the concentration-action curve for adrenalin was determined by the normally distributed variation of reaction thresholds to the drug among the susceptible 'units' or 'elements' of the uterus muscle. Moreover, Gaddum found a linear relation between the concentrations of adrenalin and its antagonist ergotamine (an alkaloid from ergot of rye) that together produced a certain intensity of muscle contraction. Further tests excluded the possibility of a chemical reaction between adrenalin and ergotamine. Also, ergotamine was not sufficiently concentrated in the uterus muscle cells to cause a measurable decrease of its concentration in the saline fluid bathing the uterus muscle preparations. This suggested merely the adsorption of ergotamine to the muscle tissue. Concerning the mode of action, Gaddum concluded:

> It is assumed that there is an area in the muscle on which adrenalin must act and that a fraction of this area is blocked by ergotamine so that in any given case the concentration of adrenalin must be increased in a certain proportion to produce the same effect ... The simplest theory that can be suggested to account for these results entails a number of elements of varying threshold to adrenalin and of varying accessibility to ergotamine.[25]

With these considerations, Gaddum introduced a concept of competitive antagonism at receptors and of receptor blockage.

Well aware of Cushny's and Gaddum's work on drug antagonism, in 1926 Clark studied the antagonistic action of acetylcholine and atropine, again using isolated ventricular strips from the frog heart and the *rectus abdominis* muscle of the frog as his experimental models.[26] In the frog heart preparations, atropine reduced the inhibition produced by acetylcholine; and in the *rectus abdominis* preparations, it reduced the contractions caused by acetylcholine. When constant effects, for example a 50 per cent reduction of the response of the heart muscle, were produced with both drugs present, Clark found the relation:

$$\text{Concentration Acetylcholine}/\text{Concentration Atropine} = \text{constant}.$$

Substituting this for x into the concentration-action formula that he had developed earlier for acetylcholine, $Kx = y/100 - y$, Clark arrived at:

$$K\text{Conc. ACh.}/\text{Conc. Atr.} = y/100 - y.$$

To Clark's satisfaction, figures calculated from this formula were consistent with the data obtained from his experiments on the antagonism between acetylcholine and atropine on the frog heart preparations. A similar formula expressed the action of the two drugs on the *rectus abdominis* muscle. Clark also found that atropine continued to exert its antagonistic action on acetylcholine even after the atropine solution had been repeatedly washed out of the heart preparations. This meant that atropine was somehow fixed by the heart cells. Moreover, as he had shown earlier in contradiction to Straub's theory, the effect of acetylcholine was not dependent upon the amount of this drug entering the muscle cells. Putting these observations together, Clark suggested that the antagonism between acetylcholine and atropine on the heart resembled the antagonism between oxygen and carbon monoxide on haemoglobin, with one important difference. While oxygen and carbon monoxide displaced each other at the haemoglobin molecule, such displacement did not seem to occur between acetylcholine and atropine, because the rate of recovery in an atropine-poisoned heart did not increase when acetylcholine was added in excess. Clark concluded from this that: 'Atropine and acetyl choline, therefore, appear to be attached to different receptors in the heart cells and their antagonism appears to be an antagonism of effects rather than of combination.'[27]

In essence, he thus introduced a distinction between competitive and non-competitive antagonism of drugs at receptors.

In light of Clark's study, the mechanism of drug antagonism seemed to be less simple than mere competition for the same receptors as proposed earlier by Cushny and Gaddum. Moreover, Clark did not entirely give up interpretations of drug actions as physical effects on cell membranes. For over a decade he had been interested in the influence of various ions on cardiac function.[28] At the time of his acetylcholine studies in the mid-1920s, Clark also investigated the action of potassium on the frog ventricle and found that the amount of potassium chloride fixed to the heart cells that was just sufficient to produce an action (reduction of the contractive force) was extremely small. He assumed that potassium exerted its effect through altering a 'lipoid film' on the surface of the heart cells, but he did not rule out the possibility of a chemical reaction between the potassium ions and the heart muscle.[29]

In 1926 Loewi and Navratil published their observation that extracts of frog heart tissue rapidly destroyed acetylcholine, presumably through an enzyme of the esterase type.[30] Clark confirmed this in his own experiments and demonstrated that the action of acetylcholine could be quickly abolished by washing the drug out of the ventricle strip. This indicated that acetylcholine acted merely on the surface of the heart cells. From these findings Clark concluded:

My experiments suggest that at least two independent processes occur when acetyl choline is brought in contact with tissues: firstly, an adsorption and destruction of the drug by the tissues, and secondly, a reaction between the drug and certain specific receptors. The latter process which produces the specific action is probably a reaction with receptors on the surface of the cells.[31]

Thus, by the late 1920s, a quantitative relation between the occupancy of receptors on the cell surface by a drug (agonist), the relative proportion of this occupancy compared with that by an antagonistic drug, and the observable pharmacological action had been established. In 1926 Clark had succeeded Cushny as the chair of materia medica at the University of Edinburgh. Six years later, he gave a series of invited lectures at University College London. Based on these lectures and a perusal of the recent research literature, Clark provided a critical synthesis of pharmacological theories in his monograph of 1933, *The Mode of Action of Drugs on Cells*. In this context he clearly formulated what is now known as the 'receptor occupancy theory': 'In many cases the shapes of the [drug concentration-action] curves obtained can be interpreted rationally on the assumption that the drug reacts with receptors, and that the action produced is directly proportional to the number of receptors occupied.'[32]

A few years later the publishers of the series *Heffter's Handbuch der Experimentellen Pharmakologie* contracted Clark as author of a volume on *General Pharmacology*, that is, a work outlining the theoretical foundations of the whole subject. In this work, published in 1937, Clark distinguished two basic types of drug action on cells. One type, exemplified by the action of phenol (formalin) on bacteria and colchicine on body cells, involved the adsorption of the drug on the cell surface, diffusion of the drug through the membrane into the interior of the cell, changes in the protein structure of the cell or its nucleus, followed by a biological response, such as cell death. The other type of drug action represented the action of acetylcholine and adrenalin. It involved the rapid occupation of specific receptors on the cell surfaces, which caused some modification of the function of the cell. Clark compared this with the action of certain poisons on enzymes, which changed enzyme function, although they reacted only with a small active group of the large enzyme molecule.[33]

Clark's *General Pharmacology* also considered a recently published communication by Gaddum on the mechanism of drug antagonism. In late 1933 Gaddum had left the National Institute of Medical Research to take up the chair of pharmacology at the University of Cairo. While his six-year period at the National Institute had been fruitful, he apparently felt that his publications were seen more as outputs of Dale's department than his own research achievements.[34] After only eighteen months in Cairo, Gaddum was appointed professor of pharmacology at University College London. Two years later he published an influential paper on the mechanism of drug antagonism in which he suggested that antagonistic drugs acted by competing with the agonist for the same receptors and by 'inertly blocking them up'.[35] Clark generally approved of this interpretation. As he emphasized in the conclusions of his book, the most interesting pharmacological problem of the time was that of the extraordinary specificity of action of certain drugs, and the assumption of specific drug receptors helped to explain not only this phenomenon but also that of antagonisms between drugs that were chemically very similar. While the agonist produced a biological effect upon combining with the receptors, the antagonist occupied them *without* producing an action. Clark now turned his back on Straub's 'poison-potential theory', stating that it had 'the disadvantage of assuming processes unknown in physical chemistry' and that the facts on which it had been based could be explained in other ways.[36] Clark had also made this statement at a scientific debate organized by the Royal Society of London in November 1936. However, Straub, who also spoke at this meeting, continued to maintain that the recent findings on the action of acetylcholine and adrenalin could also be accommodated by his 'poison-potential theory'.[37] As the debate made clear, there was no consensus among the scientific community of pharmacologists that the receptor theory was superior to other explanations of the mode of action of drugs. However, through the work of Clark and Gaddum, the receptor concept had gained significant ground by the outbreak of the Second World War.

While Clark's contributions to theoretical pharmacology were, in retrospect, his most important achievements and those that gained him international recognition, he was also a successful writer on the therapeutic aspects of the field. His textbook *Applied Pharmacology*, which had appeared in its first edition when he was still at University College London (1923), went through its seventh edition in 1940, and was translated into Spanish (1929) and Chinese (1935).[38] Clark's general approach, which was very well received, was 'to give an account of the direct scientific evidence for the therapeutic action of the more important drugs, and to demonstrate the importance of this knowledge in the clinical application of drugs'.[39] He deplored the then existing gap between 'the science of pharmacology' and 'the art of therapeutics', both in the teaching of medical students and in research. In order to bridge the gap between the results of animal experiments and clinical findings, he tried to illustrate the action of drugs wherever possible through observations on patients.[40] In 1931 Clark became a fellow of the Royal Society, and from 1934 he served on the Medical Research Council. In line with his emphasis on scientific evidence for therapeutic efficacy Clark advocated legal restrictions of the trade in 'secret remedies' (that is, proprietary medicines of undisclosed composition) and published a monograph in 1938 on this issue in the popular political series *Fact*.[41]

During the war, the chief protagonists of the two rival pharmacological theories, Clark and Straub, both died. Clark, who had served in the army as an adviser on protection against potential gas attacks, and twice went to France in this capacity, died unexpectedly at home in Edinburgh in the summer of 1941, at the age of 55, following abdominal surgery for suspected intestinal obstruction.[42] Straub, who had held the Munich chair of pharmacology since 1923, died at the age of 70 in the autumn of 1944. Interest in the theory of drug action however continued throughout the war.

In September 1943 the Faraday Society in London organized a discussion on the topic 'Modes of Drug Action', the first such conference since the death of Clark. Gaddum, who had succeeded Clark as the Edinburgh chair of materia medica in 1942,[43] was among the speakers, as was Harry Raymond Ing (1899–1974), then reader in pharmacological chemistry at University College London.[44] In the opening address of the meeting, Dale expressed his very critical view of the drug receptor theory, an attitude that can be traced back to the time immediately after Langley's formulation of the concept (see Chapters 2 and 4). Now Dale expressed his continuing scepticism with reference to Ing's paper:

> Dr. Ing alludes to the remarkable specificities of certain chemicals for physiologically [sic] effector cells – nerve, muscle and gland cells – and the association of these with innervation by different parts of the autonomic nervous system – surely one of the most fascinating of pharmacological mysteries, and one which remains a mystery, even with our present knowledge of the physiological intervention of adrenaline and acetylcholine in the transmission of different nervous effects. It is a mere statement of fact to say that the action of adrenaline picks out certain such effector-cells and leaves others unaffected; it is a simple deduction that the affected cells have a special affinity of some kind for adrenaline; but I doubt whether the attribution to such cells of 'adrenaline-receptors' does more than re-state this deduction in another form.[45]

Moreover, Dale voiced the same sentiment in connection with the main topic of Ing's presentation, that is, the relationship between the chemical structure of drugs and their biological effects. As Dale pointed out, tetramethylammonium salts had a selective stimulant action which closely resembled that of the plant alkaloids nicotine, cytisine and lobeline, but the latter three were neither 'onium' salts nor were they very similar to each other in their chemical structure. What then, he asked, was the common property underlying their common effect? He continued: 'To say that they all have affinity for the same chemoreceptors is merely to restate the observed facts or, if it means more, to go without warrant beyond them.'[46]

Despite the considerable development of the receptor theory since the time of Ehrlich and Langley, these doubts, expressed at a high-profile meeting by Dale, illustrate the controversial status it still held in the 1940s. Given Dale's prominent role in the scientific community at that time – he had received the Nobel Prize in 1936 – his sceptical comments demanded attention (see Chapter 4).

Ing, in his conference paper, acknowledged the receptor concept as a useful 'intellectual link' between the concepts of drug structure and drug action and as a means of explaining the phenomenon of antagonistic action between structurally similar substances. Yet, while he employed the receptor concept in his discussion of structure-activity relations, he also had to admit that there was still 'complete ignorance' about the actual chemical nature of receptors, and that the receptor theory only 'pushed back' the problem of pharmacological action, because it failed to explain how the drug-receptor combination produced a physiological effect. In fact, Ing is said to have maintained a rather cautious attitude towards the receptor theory throughout his working life.[47]

At the same Faraday Society conference, Gaddum, almost in defiance of Dale's scepticism, presented his interpretation of competitive antagonism between drugs on the basis of the receptor theory. Obviously, Gaddum's recent appointment as Clark's successor in the Edinburgh chair had brought about his final emancipation from Dale. His new status was also reflected in the conference organization. While Dale gave the general introductory address, Gaddum introduced a separate section of the conference. From the start of his presentation Gaddum paid tribute to his predecessor in Edinburgh: Clark's books, *The Mode of Action of Drugs on Cells* and *General Pharmacology*, he claimed, had been 'a great source of inspiration and information to all those who are interested in pharmacology'. Moreover, like Clark before him, Gaddum supported the receptor concept by drawing comparisons with findings in the enzyme research of the 1920s and 1930s, which had described competition between various chemicals and substrates at enzyme molecules.[48]

Modifications of the receptor occupancy theory

During the 1940s the concept of competitive antagonism, especially competitive *inhibition*, gained new support in the context of research on the antibiotic properties of sulfonamides.[49] In particular, it was demonstrated that sulphanilamide inhibited the growth of bacteria by competing with the metabolite para-aminobenzoic acid (PABA) for an active site on a bacterial enzyme.[50] Subsequent research on various PABA derivates revealed however the paradoxical phenomenon that some of these chemical substances could have both metabolite-like (that is, growth stimulating) and inhibitory effects on the same microorganism, depending on the experimental conditions.

In the early 1950s, the Dutch pharmacologist Everhardus Jacobus Ariëns investigated this problem, showing that similar 'dual action' applied also to some muscle paralysing agents and sympathomimetics. In a series of derivatives of such drugs, some derivatives produced the opposite effect of the drug (for example, muscle contraction instead of muscle paralysis). Other derivatives could display either effect depending on the experimental setting. Ariëns realized that Clark's receptor theory failed to account for such 'dual action'. Clark had merely assumed either pure agonists that produced a full effect on occupying their receptor, or pure antagonists that produced no effect at all when combining with the same receptor. Ariëns extended this theory by introducing the 'intrinsic activity' of a drug as a new factor that determined its effect. While a drug's 'affinity' determined how many drug-receptor combinations were formed, its 'intrinsic activity' determined the potency of the biological effect that was initiated by the forming of drug-receptor complexes. In a general sense, this distinction reflected Ehrlich's differentiation of haptophore (binding) and toxophore (poisoning) groups in toxins (see Chapter 1). Both 'affinity' and 'intrinsic activity' contributed to the overall pharmacological action of a drug. In this way the apparently anomalous cases of 'dual effects' could be explained. Such substances had 'intermediate intrinsic activities'. If, for example, a PABA derivative had less intrinsic activity than the metabolite PABA, it acted as a competitive inhibitor of bacterial growth in the presence of the metabolite. When the metabolite was absent, however, the PABA derivative's own intrinsic activity showed, that is, it then acted as a growth factor on the bacteria. In general terms, therefore, the biological effect of a drug had to be described not only as a function of its dose (or concentration) and its affinity to the receptor, but also as a function of the amount of effect it was able to produce upon combining with the receptor.[51]

A colleague of Gaddum at the Edinburgh department of pharmacology, Robert P. Stephenson (1925–2004), proposed a similar extension of the receptor theory in 1956. Stephenson pointed to the variety of slopes of drug concentration-action curves that had been found since Clark's work on acetylcholine in the mid-1920s. This meant that Clark's formula, $Kx = y/100 - y$, for a monomolecular reaction between drug and receptor, had to be modified by postulating that one drug molecule combined with either less or more than one receptor. Such a proposition, however, seemed unlikely, and Stephenson therefore challenged Clark's assumption that the proportion of the receptors occupied by a drug was directly proportional to the biological response of the tissue. Stephenson formulated the following three hypotheses:

(1) A maximum effect can be produced by an agonist when occupying only a small proportion of the receptors.
(2) The response is not linearly proportional to the number of receptors occupied.
(3) Different drugs may have varying capacities to initiate a response and consequently occupy different proportions of the receptors when producing equal responses. This property will be referred to as the efficacy of the drug.[52]

Actually, the first of these three hypotheses was not entirely at variance with the views of Clark, who had indicated the possibility that a drug produced its maximum action before having occupied all of its available receptors.[53] However, Clark explained differences in potency between structurally similar drugs with differences in affinity to the same receptor.[54] The response of a cell to receptor occupancy was believed to be of an 'all or none' type. Stephenson, by contrast, introduced the concept of the 'efficacy' of a drug. It could vary from zero (no response, receptor blockage) to some high positive value. This led to a more differentiated understanding of drug antagonism. Stephenson called those drugs with a low efficacy, and thus with properties lying between those of (pure) agonists and antagonists, 'partial agonists'. The partial agonists in Stephenson's terminology more or less matched Ariëns's notion of drugs with 'intermediate intrinsic activity', which could act as agonists or as antagonists, depending on the experimental conditions.

Moreover, Stephenson named receptors that were still unoccupied even after a drug had produced its full effect, 'spare receptors'. This concept helped to explain apparent transitions from competitive antagonism to non-competitive antagonism in some drugs that blocked the effects of histamine. As long as a sufficient number of receptors remained unoccupied by the blocking drug, an increase in the concentration of the agonist could 'surmount' the blockage through combinations with those spare receptors. When the antagonist drug had occupied all the spare receptors as well, then the blockage became 'insurmountable'.[55]

Stephenson had found evidence for his views in experiments with a series of alkyl-trimethylammonium salts, which he tested on the isolated guinea-pig ileum and whose concentration-action curves largely matched those that he calculated, on the basis of his theoretical assumptions, from the affinities and efficacies of these compounds. Even so, he remained concerned about potential criticism from those who continued to be sceptical of the receptor concept as such: 'The approach to study the action of drugs used in this paper is not universally popular among pharmacologists; some, indeed, despise discussions in terms of receptors...'[56]

However, the contributions of Ariëns and Stephenson led the way towards ever more sophisticated experimental investigations and mathematical treatments of the action of drugs on the – still hypothetical – receptors. For example, in 1957, Bernard Katz (1911–2003) of the department of biophysics at University College London, who had been a student and assistant of A.V. Hill, explored the phenomenon of 'desensitization' of acetylcholine receptors using a new method of microapplication of drugs to motor nerve endings in frog muscle. On the basis of this work he postulated two states or forms of the receptor, an 'effective' and a 'refractory' form, with which acetylcholine or similar drugs could combine. If a drug's affinity to the refractory form was much higher than its affinity to the effective form, it produced profound desensitization and had only little effect on the muscle.[57]

In 1961 William D.M. Paton, professor of pharmacology at the University of Oxford, proposed a rate theory of drug-receptor combination. This theory was based on experiments performed on the isolated ileum of the guinea pig, using acetylcholine and histamine as agonists, hyoscine, mepyramine and atropine as antagonists, and alkyl-trimethylammonium compounds as partial agonists. It rested on the assumption that the biological effect produced by an agonist drug was proportional to the *rate of combinations* formed between drug molecules and receptors, rather than to the proportion of receptors occupied by the drug. If a drug had a high dissociation rate, that is, if the drug molecule left its receptor quickly to form another transient combination with the next receptor, this drug was a strong agonist or stimulant. If its dissociation rate was low, that is, if its molecules were bound for a longer period to the same receptors, thus preventing other drug molecules from forming combinations with these receptors, it acted as an antagonist. If the dissociation rate of a drug was moderate, then it was a 'partial agonist' in the sense of Stephenson.[58] Paton coined a memorable musical analogy to illustrate the difference between drug-receptor interaction according to Clark's receptor occupancy theory and his new rate theory. The occupancy theory of drug action resembled playing on an organ: a sound is produced as long as a key is being pressed. Drug action, according to Paton's new theory, was like playing a piano: it makes a sound when a key is struck, but the sound is transient, and as long as the key is being held down the hammer cannot operate again.[59]

Paton's rate theory did not win acceptance over the receptor occupancy theory, which was further developed and refined. However it reflected the considerable momentum that work on drug receptor theory had gained by the start of the 1960s. Ariëns, who had founded the pharmacological institute at the Catholic University of Nijmwegen in 1951, edited a two-volume handbook on *Molecular Pharmacology*. Published in 1964, the first volume contained over 400 pages of detailed, state-of-the-art discussions of receptor theory, and the second volume included various applications, such as the physiology of olfaction, chemotherapy of cancer and enzymology.[60] Yet, despite the many research results that been obtained within the receptor paradigm, receptors were still hypothetical entities. As D.K. de Jongh, professor of pharmacology in Amsterdam, put it in his introduction to Ariëns's handbook:

> To most of the modern pharmacologists the receptor is like a beautiful but remote lady. He has written her many a letter and quite often she has answered the letters. From these answers the pharmacologist has built himself an image of this fair lady. He cannot, however, truly claim ever to have seen her, although one day he may do so.[61]

Conclusions

This chapter has shown how the concept of specific drug receptors, introduced by Langley and Ehrlich at the beginning of the twentieth century, evolved into an increasingly complex theory of drug-receptor interaction. Although prominent scientists, most notably Sir Henry Dale, remained sceptical about the heuristic value of the receptor concept, and although this concept competed with other interpretations of drug action, especially the 'poison-potential theory' of Walther Straub, it slowly developed into a central tenet of pharmacology. Crucial to this development was a quantitative and mathematical approach to the study of drug action, particularly through the analysis of concentration-effect curves. While this quantitative approach had been used by Hill as early as 1909, it was only during the 1920s that it was more fully exploited in the work of Clark on acetylcholine and Gaddum on adrenalin. The contemporary interest in chemical transmission of nerve impulses to effector cells formed a background to this kind of pharmacological research. Increasingly, the receptor concept became 'quantified'.

A network of receptor theorists came from the Cambridge school of physiology where Langley had taught several important contributors to the field, including Hill, Clark and Gaddum. In the 1920s and 1930s, University College London, with Clark and then Gaddum, emerged as a first centre for receptor pharmacology. With the appointment of Clark as Cushny's successor in 1926, and then Gaddum as Clark's successor in 1942, to the Edinburgh chair of materia medica, two advocates of the receptor theory had obtained key positions in British pharmacology. With the work of Stephenson in the 1950s, Edinburgh remained at the forefront of receptor research. Nevertheless, the development of receptor theory between 1910 and 1960 was driven by a still relatively small, though prestigious, group of pharmacologists. Their work contributed to a theoretical foundation for pharmacology, as documented by Clark's book *General Pharmacology* of 1937 and the *Molecular Pharmacology* volumes edited by Ariëns in 1964.[62] The concepts of receptor occupancy, competitive and non-competitive antagonism, and partial agonists became central to pharmacological reasoning. However, until the 1950s, receptor research was chiefly a pursuit of basic academic pharmacology with strong links to physiology. The potential of the receptor concept for the development of new pharmaceuticals became only gradually clear from the late 1950s, with work on receptor-subtype specific drugs. The following chapter will discuss this important new phase in the history of the receptor idea.

6

The Dual Adrenalin Receptor Theory of Raymond P. Ahlquist (1914–83) and its Application in Drug Development between 1950 and 1970

In the previous chapters we have seen that the early introduction of the receptor concept by Langley and Ehrlich was a very hypothetical approach and was considered only as one of a number of options available to clarify the problem of drug binding. Consequently, its fate remained uncertain within the scientific community of pharmacologists during the first half of the twentieth century. The most important representatives of the discipline propagated competing alternative theories as well as alternative research strands. In spite of this, the receptor concept survived, particularly through the development of quantitative theories of drug action, which achieved a clear breakthrough in the 1930s with Alfred Joseph Clark's work on drug binding to cells. After Clark's early death, scientists such as Gaddum, Ariëns and Stephenson deepened the research on the mechanism of drug binding to receptors (see Chapter 5).

By the late 1940s, there was a group of pharmacologists devoted to research on receptors, but these individuals represented a minority of researchers in pharmacology and the majority remained committed to traditional approaches, which were deeply rooted in the educational and research programmes of the field. Cushny, Dixon and Straub had died, but Sir Henry Hallett Dale, whose reputation was further enhanced as a Nobel Prize winner, now dominated research policy in the UK and acted internationally as a spokesman for this majority.

Those who propagated the receptor concept were still on the defensive when the concept was further advanced by new investigations that profited from the post-war climate. The war against Hitler's Germany and Japan had led to new scientific developments in the allied countries in diverse areas of research. On the one hand, these new developments could be seen as a threat and danger to civilization. On the other hand, war-related developments in scientific knowledge, as well as of instruments and tools, did promote, to a certain extent, further investigations in the civil sector. Science, with all its ambiguities, was seen as the key to solving many problems, and medicine regarded itself as a largely unquestioned explanatory authority in all matters related to health and disease.[1]

The pharmacologist Raymond P. Ahlquist (1914–83) is in some way representative of this situation, and he gave decisive new impetus to research on receptors. This chapter will therefore deal above all with *his* work and how it contributed to the gradual breakthrough of the receptor idea in pharmacology after approximately 1960. First, we will concentrate on the development of his concept. Second, we will deal with the publication of his work and the reaction of the scientific community to his approach. Third, we will focus on the reasons why it took time for Ahlquist's concept to be finally accepted. Fourth, we will deal with the breakthrough of Ahlquist's dual adrenoceptor theory and finally we will analyse this whole subject and draw conclusions.

Raymond P. Ahlquist and the development of the 'dual adrenoceptor theory'

Ahlquist was born in 1914 in Missoula, Montana (USA), the son of Swedish immigrants. He studied pharmacy at the University of Washington in Seattle, receiving his BSc in 1935. Under the supervision of pharmacologist J.M. Dille, he earned his Master's in 1937 and his doctoral degree in 1940. Between 1940 and 1944, he was assistant professor of pharmacology and pharmacognosy at South Dakota State College in Brookings (USA). In 1944, he started work as an assistant professor of pharmacology at the Medical College of Georgia in Augusta. In 1948, he became professor and chairman of the department of pharmacology as successor to Robert A. Woodbury (1904–2001), who had been head of the department since 1943. Ahlquist held this position until 1963, and then again between 1970 and 1977. In 1977, he acquired the Charbonnier professorship. Ahlquist remained in Augusta until his death in 1983.[2]

During his time at South Dakota State College, Ahlquist became increasingly interested in research on the functional aspects of the autonomic nervous system. As we have seen in the previous chapters, it was well known that specific substances influenced the activity of this system and the challenge was to investigate in detail exactly which substances transfer which information to the end organ or the nerves. The final aim was to understand the autonomic nervous system in more depth in order to use its functions for the sake of therapy or to combat unwanted symptoms in patients. The work of various pharmacologists had already shown that there were chances of therapeutic application.

With such aims in mind, Ahlquist participated in a project to domesticate *Ephedra sinica*, a plant which was the main source of ephedrine (a sympathomimetic amine) during the 1940s. Although the shortage of ephedrine was quickly overcome by successful synthesis of the substance, Ahlquist continued to be interested in sympathomimetic drugs – substances with an adrenalin-like effect. But in the course of experiments, which were performed with only 'relatively crude'[3] instrumentation, Ahlquist soon faced the crucial problem of how exactly drug binding to cells took place, since there was no generally accepted theory at hand. Sympathomimetic substances had contradictory effects on the circulation, the heart and the vascular system of laboratory animals. This made it impossible to assign a specific profile of effects to a specific amine and therewith to a specific chemical structure. Especially troubling was the finding of 'reversal' effects of the respective substances, for example, that octin, a sympathomimetic amine, on the one hand produced a rise of the heart rate as a 'typical' sympathomimetic effect, but on the other hand caused a depressor response. Also, the substances showed different profiles of effects in different tissues and organs. It was basically the same problem that had been addressed by, among others, Barger and Dale about 30 years earlier, in 1910.[4]

Ahlquist's studies after his arrival at the Medical College of Georgia in Augusta provided the answer to this problem. There he worked with Robert A. Woodbury (then head of the small pharmacology department), who had already performed studies on the effect of amines on the cardiovascular system and especially on the blood vessels of the uterus.[5] During his collaboration with Ahlquist, Woodbury tried to find a drug that was able to combat the symptoms of dysmenorrhoea, and it was the search for a sympathomimetic substance that would effect muscle relaxation of the uterus, which led Ahlquist to a successful contribution to receptor research.[6] Ahlquist and Woodbury's research work greatly benefited from new experimental tools that had been developed during the war by the physiologist William F. Hamilton (1893–1964), then head of the physiology department of the Medical College. New instruments allowed a more accurate measurement of blood flow and blood pressure than had previously been possible, enabling investigators to distinguish and examine diverse parameters that influenced blood pressure. These parameters were cardiac functions and disturbances, but there were others which had their origins in the vessels themselves. For example, the Hamilton hypodermic manometer allowed very sensitive, optically recorded measurements of blood pressure via deflections of a small copper membrane. Electromanometers, which were introduced after the war, relied on the deflection of a metal membrane, which was recorded electrically. The latter was an invention of the aircraft industry.[7]

In Augusta, aided by these new tools, backed by inspiring collaborators, and with practical aims in mind, Ahlquist's first step to finding a therapeutic application was to try to discover the exact profile of the effect of diverse sympathomimetic substances. The resulting pragmatic preliminary studies turned out to be the basis for the development of his dual adrenalin receptor theory. In the following, we will describe Ahlquist's experimental and theoretical approach.

He concentrated mainly on six substances – noradrenaline, cobefrine, adrenalin, levo-adrenaline, methyladrenaline and isoprenaline. These six substances were tested on 23 different tissues of research animals – cats, dogs, rats and rabbits. These tissues belonged to different organs and five organ systems were tested in total. These were the cardiovascular system (renal, mesenteric and femoral vascular beds, isolated and intact heart[8]), the intestine (isolated and intact ileum), the uterus (isolated and intact uterus), the ureter (intact ureter) and the eye (intact *dilator pupillae* and nictitating membrane).[9] It is immediately clear that new tools allowed Ahlquist to investigate the heart and the vascular system independently and to differentiate the effects of the substances in question.

The instruments used by Ahlquist demonstrate the importance of postwar technological advances on medical research in general and on his pharmacological work in particular. The blood flow and the vasomotor resistance (that is, the mechanical resistance of vessels reducing the speed of blood flow) of the vascular beds were measured and recorded with modern flowmeters, which were introduced into the arterial inflow or the venous outflow. For investigations on isolated hearts, Ahlquist used an optical lever system that was made out of a former bomb-sight. For recording the general arterial pressure, Ahlquist used not only the older mercury manometers but also Hamilton's manometer and, with many dogs, a high-frequency Hamilton manometer. The Hamilton manometer was also applied in the investigations of the intestine and the uterus. Experimental results on the eye and ureter were recorded with a string and lever system or a drop counter. In the investigation of the intact animal, the substance in question was injected into a vein or an artery; in experiments on isolated organs, it was applied directly or the organs were perfused.[10]

Ahlquist summarized the results of all experiments in a table, ranking the effect of all substances for each organ system in terms of potency and function. Potency meant the strength of effect that was related to the function of the organ, for example, vasoconstriction or vasodilation, excitatory or inhibitory effect. His analysis produced the remarkable result that the effect of the six amines examined could be reduced to two orders of potency for specific functions. The first order was:

1. Levo-adrenaline, 2. Adrenalin, 3. Noradrenaline, 4. Cobefrine, 5. Methyladrenaline, 6. Isoprenaline.

With decreasing potency these first-order substances induced in all animals vasoconstriction, excitation of the uterus, nictitating membrane, *dilator pupillae* and ureter. Also, they had an inhibitory effect on the intestine. The second order was:

1. Isoprenaline, 2. Levo-adrenaline, 3. Methyladrenaline, 4. Adrenalin, 5. Cobefrine, 6. Noradrenaline.

With decreasing potency these second-order substances induced vasodilation and myocardial excitation in all animals. Also, they had an inhibitory effect on the contraction of the uterus.[11]

Ahlquist concluded that the reason for this specification of substances could only be found in relation to the organism's structures – organs, tissues and cells. Thus, he was led to the use of the receptor idea to explain the phenomenon. Ahlquist came to the conclusion that there must be two different binding sites for substances at the cells, two kinds of adrenergic receptors, which he termed α- and β-receptors. The α-receptors would be related to the first order of amines and therewith to 'most of the excitatory functions (vasoconstriction, and stimulation of the uterus, nicitating membrane, ureter and dilator pupillae) and one important inhibitory function (intestinal relaxation)'. The β-receptors would be related to 'most of the inhibitory functions (vasodilation, and inhibition of the uterine and bronchial musculature) and one excitatory function (myocardial stimulation)'.[12] With this theory, Ahlquist not only added one new receptor theory to the already existing ones, he also brought discussions about receptors to a new level. While Clark, Gaddum, Ariëns, Stephenson and others focused their investigations on the mode of action of 'the' receptor, Ahlquist now, to a certain extent, went back to Langley's and Ehrlich's theories. Both of them had conjectured that there were different kinds of receptors, the latter had even speculated about a microcosm of receptors. Ahlquist, in a way, confirmed their speculations on the basis of exact experiments with amines, thus opening a door to the idea that there were different types of receptors. This was a prerequisite to trusting the further development of elaborate receptor theories and it increased the chances of integrating the receptor idea into pharmacological research concepts.

Publication, and the reaction to Ahlquist's approach

Ahlquist published his results in 1948 – but only with difficulty. The basic problem was not only that he ventured into the contentious field of receptors but also that he launched an attack on the well-known and widely accepted theory of Walter Bradford Cannon (1871–1945) and his collaborator Arturo Rosenblueth (1900–70) from Harvard Medical School. These two researchers had explained all the ambiguities regarding the effect of adrenergic substances with a baroque theory deduced from investigations with adrenalin. Two mediator substances in the cells, namely the inhibitory 'sympathin I' and the excitatory 'sympathin E', would be decisive for adrenalin and pharmacologically-related substances and their divergent profile of effects: there was one receptor which could release two different 'sympathins'.[13] Ahlquist's criticism of this theory was risky; Cannon was considered the most important American physiologist of the time and his assistant Arturo Rosenblueth, the son of a Hungarian immigrant and a Mexican-American woman, was seen as a gifted and talented young scientist, even if he did have a tendency towards arrogance. Rosenblueth was extremely productive and undoubtedly boosted the already high reputation of Cannon's physiological laboratory at Harvard University, which had served to train many promising young researchers.[14]

Ahlquist launched his attack as was commensurate with his personality, clearly and plainly. His criticism was unavoidable as it was implicit within his findings: there were two types of receptors, each of them having a specific profile of effects, but both having basically excitatory as well as inhibitory functions. Also, according to Ahlquist, epinephrine (adrenalin) was the most active substance on both receptor types, having therefore inhibitory as well as excitatory properties. This meant that it was not feasible to argue for two mediator substances E and I, one being 'excitatory' and the other 'inhibitory'. Furthermore, the receptor type was most decisive for the effect of the substance and not the mediator itself.[15]

The consequence of Ahlquist's criticism was that his paper was rejected by the *Journal of Pharmacology and Experimental Therapeutics*. The argument of the editor of the journal was that Ahlquist had condemned Cannon and Rosenblueth's theory, which was seen at this time as a ' "law" of physiology'.[16] It was his friend and colleague Hamilton who helped Ahlquist. His influence as an editor enabled publication of the paper in the *American Journal of Physiology*. With hindsight, it could be seen as something of a curiosity that a paper which spoke against the 'law of physiology' could be published in a journal of physiology, but in fact it was not a surprising outcome, as it corresponded with William Hamilton's scientific policy and his decision to support Ahlquist. Hamilton was keen to support experimental work in physiology and pharmacology at the Medical College in Georgia. Furthermore, he promoted collaboration between physiologists and pharmacologists. The staff of both departments spent lunchtimes together and discussed their findings and theories frankly. It was an open-minded atmosphere that surely contributed to the development of Ahlquist's theories. Hamilton had equipped Ahlquist's laboratory with his latest experimental gadgets. He promoted original and critical work in pharmacology in general and above all the creative work of Ahlquist. They were friends who together 'turned out many fine post-doctoral trainees'.[17] Hamilton also had close contact with the clinicians in Georgia, and it might be that he saw the practical implications of Ahlquist's work. Against this background, Hamilton's support for the publication of Ahlquist's decisive paper in 1948 can be seen as the first step in a larger strategy that was to have major consequences.[18]

Publication overcame the first of Ahlquist's problems, but subsequent steps, namely the public acceptance of his theory, the recognition of the full impact of his theory, and finally the awarding of honours to Ahlquist were also fraught with difficulty. It took ten years for his 1948 paper to be brought into scientific debate and its important practical implications were first considered only in the 1960s. Ahlquist never received what many of his friends and colleagues expected, the Nobel Prize in Physiology or Medicine. It is tempting to explain the late breakthrough of the dual adrenoceptor concept primarily with reference to the scientific disputes about how it fitted with then current knowledge about drug binding.[19] However, if we look at the post-1948 story in more detail, at the steps taken by Ahlquist and what happened to him, things become more complicated than that.

Although Ahlquist received no immediate positive feedback on his publication, he continued to work on amines and especially adrenalin (epinephrine) after 1948. One important idea was to block active bodily substances via the application of inhibitory drugs in order to gain more knowledge about the transfer of information and also about the receptors. In the next ten years Ahlquist experimented with adrenergic blocking agents to verify details of the profile of effects of the α- and β-receptors. In 1957, for example, together with his first post-doctoral trainee Bernhard Levy, he investigated 22 compounds and their effects on the depressor response of epinephrine in anaesthetized dogs, which had been pre-treated with two α-blocking substances, phenoxybenzamine and dibenamine. Because of the blocking substances, epinephrine had a depressor effect on the vascular system. Levy and Ahlquist found that only two substances, namely ephedrine and methoxamine, could reverse the depressor effect of epinephrine to a pressor effect, and they speculated that these substances would be able to 'displace the adrenergic blocking agent from its site of action'.[20] Ahlquist's collaboration with Levy was very successful. The latter became Ahlquist's research colleague, and they spent 16 years together investigating β-receptors. Levy finally suggested that there were three β-receptors: a cardiac one, one located at the blood vessels, and one at the bronchi.[21]

In the course of his further work on the sympathetic system, Ahlquist participated not only in research on α-blocking agents, but also promoted research on the characterization of β-blocking agents, which became more and more important in the decades to follow.[22] Alpha-blockers were well known,[23] but there was not much knowledge of β-blocking agents, which were, as a consequence, also under investigation by other research groups. One of these groups worked in the Lilly Research Laboratories in Indianapolis (USA). Eli Lilly and Company had already sponsored Woodbury and Ahlquist's research in Augusta, and the company's laboratories continued to pursue the same area in the search for a substance with bronchodilator activity.[24] One of the compounds, namely dichloroisoprenaline (DCI, compound No. 20522), synthesized by J. Mills from the organic chemistry division of the company, showed inhibition of inhibitory actions of adrenalin, among them broncho- and vasodilation. The results were published by C.E. Powell and Irwin Slater in 1958. They remarked that the results of their experiments would have 'special relevance when considered in terms of drug-receptor theory', but they did not relate their findings to Ahlquist's dual adrenoreceptor concept, nor did they quote their colleague from Augusta.[25] Rather, they saw their experiments as fitting into the area of research that had been developed by Clark and then followed by Gaddum, Ariëns and Stephenson. Quoting Ariëns and Stephenson, according to Powell and Slater the effect of compound 20522 was a good 'example of the separation of affinity and intrinsic activity'. The substance combined 'with "adrenergic inhibitory sites"' but would fail 'to trigger the series of reactions that lead to typical inhibitory effects'. Furthermore, one of the most important results was that compound 20522 would 'have a reasonably long biological life'. As the substance caused a stable blockade of the adrenergic receptors, it was seen as a useful tool for further laboratory experiments on the sympathetic system.[26]

However, Ahlquist was soon brought into the game as a further step linked Powell's and Slater's blocking agent of adrenergic sites with Ahlquist's dual adrenoceptor concept. This step was taken by Neil C. Moran (1924–97), then head of the department of pharmacology (division of basic health sciences) at Emory University in Atlanta, Georgia, only 120 miles away from Augusta. He heard a presentation by Irwin Slater at the meeting of the Federation of American Societies for Experimental Biology in Chicago, Illinois, in 1957. In this lecture, Slater presented the results of his research on DCI, which was the basis of the paper published jointly with Powell one year later. Moran was a keen competitor of Ahlquist and knew Ahlquist's work and theories very well. He was struck by the idea that Slater's and Powell's results would fit with Ahlquist's theory and that the heart especially would play the central role within the sphere of activity of DCI. He even thought that DCI would be a β-receptor blocking substance. He asked Slater for a sample of DCI, which he got two weeks later.[27] Together with his colleague Marjorie E. Perkins, he then investigated in detail the effect of the substance. They used anaesthetized dogs and isolated rabbit hearts to examine the influence of DCI on the effect of different amines. In the case of the dogs, the influence of the parasympathetic nerves was removed (through vagotomy), and the different substances were applied through the femoral vein. The contractile force of the right ventricle of the heart was measured as well as the arterial pressure. The rabbit hearts were removed from freshly killed rabbits and perfused with a nutritive solution, and the substances were injected above the canulated aorta. The contractions of the heart were recorded on an oscillograph.[28] The studies conducted by Moran and Perkins show the ways in which the experimental setting of pharmacologists had changed, as the investigation of effects on the animal were complemented by investigations on the exact processes of drug binding.

The main result of the experiments in dogs and rabbits was that DCI blocked exactly those effects that had been assigned by Ahlquist to the second order of sympathetic substances. It was 'a selective blockade of most inhibitory functions' and 'cardiac positive inotropic and chronotropic effects'. DCI therefore represented a 'specific cardiac adrenergic blockade' and in the view of Moran and Perkins, this supported the 'postulate of Ahlquist (1948) that the adrenotropic inhibitory receptors and the cardiac chronotropic and inotropic adrenergic receptors are functionally identical, *i.e.*, that both are *beta* type receptors'. DCI was explicitly described as a β-blocking substance.[29] Thus Moran and his colleague related the research on blocking agents of the sympathetic system to Ahlquist's theory, introducing it to experimental basic pharmacology as an explanatory tool of research on adrenergic receptors. With a clear feeling that Ahlquist had made an important step in the research on the sympathetic system, Moran managed at least to participate in the honours Ahlquist received for his work. Like Powell and Slater, and only six months later, Moran and Perkins submitted their paper to the *Journal of Pharmacology and Experimental Therapeutics*, where it was also published in 1958.[30]

Powell and Slater quickly adopted Moran's and Perkin's view. Only a year later, they published a contribution to a symposium on catecholamines, this time with Slater as first author. The topic again was the effect of DCI, but this time, they explicitly pointed out that the substance would be a useful tool for separating α- and β-receptors. Moreover, quoting Ahlquist and Moran and Perkins, they now suggested that their data supported Ahlquist's concept of α- and β-receptors.[31]

In later publications Moran maintained his image as Ahlquist-discoverer and consistent supporter. In 1961, again with Marjorie E. Perkins, Moran published research on α-receptors in the *Journal of Pharmacology and Experimental Therapeutics*. This was a reaction to some reports that α-receptor blocking agents – like the β-blocking ones – would also block the sympathetic inotropic, strengthening cardiac action. This would have meant diminishing or playing down any differences between α- and β-receptors and devaluing Ahlquist's concept. Moran and Perkins saw their work as an extension of their experiments performed in 1958. They used the same experimental setting but now with the aim of investigating the impact of some α-blocking agents (phenoxybenzamine, phentolamine, chlorpromazine and dihydroergotamine) on the positive inotropic cardiac response. They found no evidence for any selective cardiac adrenergic properties.[32] The paper was a defence of Ahlquist's theory and a defence of their own influence in popularizing it in the scientific community. Quoting Ahlquist's paper from 1948, they stated that: 'The differentiation between compounds of this type [that is, the substances mentioned above] from DCI supports the suggestion put forth by Moran and Perkins (1958) that adrenergic blocking drugs be classified according to the terminology of adrenotropic receptors introduced by Ahlquist (1948, 1958).'[33]

Why was Ahlquist's concept not immediately accepted? Reasons and theories

(A) The scientific community of pharmacologists

Looking at the network of pharmacologists after Ahlquist had published his theory, it is clear that the acceptance of his theory faced a variety of hurdles, which can be sorted into three different major categories.

First, Ahlquist was in competition with other research groups who were also devoted to the investigation of the sympathetic system – sometimes with the same sponsors. Consequently, it was difficult to gain rapid acceptance for any concept. Powell and Slater took up Ahlquist's approach only following the example of Moran and Perkins, who supported Ahlquist because his theories underlay their own investigations. Furthermore, there was still competition from Cannon and Rosenblueth's well-known and dominant theory and the idea that adrenalin is the substance which increases blood pressure and cardiac activity but not vasodilation. In this context, Ahlquist's differentiated theory of the adrenergic system was hard to accept and understand.

Second, there were inherent problems with Ahlquist's theory itself. Other researchers contradicted it: for example, in 1946, Ulf von Euler (1905–83), claimed that noradrenaline and not adrenalin was the main mediator of the sympathetic system.[34] Further, the technical limitations of the day meant that no accurate visual image of receptors could be achieved. These limitations meant that, despite their support for the theory, neither Moran and Perkins nor Powell and Slater were particularly enthusiastic in promulgating Ahlquist's theory as the decisive event for the understanding of the sympathetic system. In their 1958 paper Moran and Perkins pointed out that the 'structural requirements for cardiac adrenergic blocking drugs are poorly defined as yet'. In this sense, Ahlquist's concept was no more than 'a useful classification of adrenergic receptors'. In their 1961 paper, with hindsight, they called their proposal to classify adrenergic blocking drugs according to Ahlquist's terminology of adrenotropic receptors only a 'suggestion'.[35] In 1959, Powell and Slater related Ahlquist's concept to their self-critical view of DCI-research:

> Thus, 20522 can be considered a useful tool for the quantitative separation of α- and β-receptor sites, but for the present it must be used

with caution for the quantitative aspects are not clear. In any event, the data seem to add support for the Ahlquist concept of α- and β-adrenergic receptors and suggest that this separation provides a useful frame of reference in discussions of sympathomimetic amines.[36]

This 'suggestion' provided clear, but hardly enthusiastic, support for Ahlquist.

Third, from 1958 until the early 1960s Ahlquist's support came from within the realm of theoretical pharmacology. Moran and Perkins as well as Powell and Slater investigated DCI mainly with the aim of fostering research on the sympathetic system as basic research in pharmacology. They did not intend to promote therapy in a more direct way, and as Slater commented retrospectively: 'We had few ideas for clinical utility and were not in a position to do detailed cardiovascular studies.'[37] Robin Shanks, who worked with Archibald D.M. Greenfield and W.E. Glover at the department of pharmacology at Queen's University Belfast, was the first to test and demonstrate the β-blocking effect of DCI in man. As a research fellow and post-doctoral trainee of Ahlquist, Shanks had attended a seminar given by Neil Moran in 1960 at the department of pharmacology and physiology in Augusta. This was the starting point of Shanks's interest in the antagonism of DCI and adrenalin, but on a purely theoretical basis. Even Shanks and his collaborators did not think about any therapeutic application of the substance. Ahlquist himself was also unable to investigate and explain the relevance of his research for practical clinical medicine.

This meant that Ahlquist's theory had only a limited impact on medicine, and it was only as a result of an initiative from the side of clinical practice that this situation changed (see below).[38] Finally, one has also to consider that there was – in spite of Hamilton – no support from the scientific community of physiologists and that there was only a small group of pharmacologists working on receptors when investigating the relationship between dose and response on the basis of A.J. Clark's concept.[39]

(B) Ahlquist's character and his own attitude towards the 'dual adrenoceptor concept'

Besides scientific debates about Ahlquist's theories, which were characterized by rivalry, Ahlquist's personality and his attitude towards his own research also influenced the fate of his theory. He could be 'brutally frank', especially in discussion on scientific matters. His open criticism of Cannon and Rosenblueth had already shown glimpses of this tendency, for which he soon became well known among his colleagues in Augusta. One day, an applicant for medical school was interviewed by Ahlquist and left with 'stars in his eyes'. Being informed of this, Ahlquist remarked 'We'll take 'em out of him in the first year.'[40] Although he could be charming, Ahlquist said what he thought, without caring much about the consequences. Speeches at Ahlquist's funeral in 1983 hinted at this. In contrast to Hamilton, who was polite, personable and obliging, Ahlquist was crusty and difficult to handle.[41] He could be 'a very private person' and 'at times abrasive'.[42] His habit of saying what immediately came to mind was not confined to students. He was also very open with his criticism when talking with his colleagues.[43] After Ahlquist attended a presentation on the treatment of hypertension with clonidin, given by a young assistant professor undergoing review by his successor Lowell Greenbaum (chair of pharmacology), Ahlquist told the young man: 'It was a nice lecture, but I don't believe a word of it.'[44]

Ahlquist wanted to promote an open, critical discussion of scientific results, which while broadly the intention of the scientific community was actually rejected by many researchers in its most radical or direct form. Ahlquist's own robust style of communication seems consequently to have affected the promotion and reputation of his receptor theory. Nonetheless, Ahlquist received many prizes and acknowledgements for his work: in 1974, he received the Hunter Memorial Award of the American Society for Clinical Pharmacology and Therapeutics; in 1976, together with Sir James Black, he received the Albert Lasker Clinical Medical Research Award; and in the same year the CIBA Award for Hypertension. Also, in these years, Ahlquist's achievements were pointed out at several international symposia on β-blockers. In 1980, he was honoured by the holding of an 'International Symposium on Adrenergic Receptor Pharmacology'. This was also the occasion of the first annual meeting of the Southeastern Pharmacology Society (SEPS), which had been founded by Lowell Greenbaum.[45] However, Ahlquist did not receive the Abel Award, one of the most important prizes in American pharmacology, and he failed to get the most important scientific award of all, the Nobel Prize. Everyone in Georgia expected him to receive it, but Ahlquist 'was a complex man who had love/hate relationships with some which no doubt, at the international level, cost him the Nobel prize'.[46] Ahlquist himself believed that one of the Nobel officials was 'a personal enemy of mine'.[47]

Diplomacy was not one of Ahlquist's strong points, and there are numerous indications that this also had an impact on his career in

Augusta. During his time as dean of the Medical School, he controlled money and research activity on the basis of sometimes rigid decision-making and made enemies as well as friends. In 1977 he was sued for not granting tenure to an individual in his department. He even lost the chair of the department and was banished to an isolated building on the MCG campus without any laboratory facilities. Becoming Charbonnier professor in this sense was not only an honour, but also a drawback for Ahlquist. Five years later, in 1982, his successor, Lowell Greenbaum, who had come from Columbia University to Augusta in 1979, offered him an office and a laboratory in the department and thereby the chance to come back. Ahlquist happily agreed, but shortly before moving to his new facilities, he died in hospital.[48]

Another feature of Ahlquist's personality was a degree of scepticism or pessimism.[49] Ahlquist was proud of his theory and was honoured for it, yet he grumbled about those honours that had been withheld from him and about not getting enough recognition for having developed his receptor concept. Ahlquist was also annoyed with himself for having worked with the first β-blocking agent without recognizing what it was – years before these substances became fashionable. His case proved 'that a prophet is without honor in his own country'. In his opinion he belonged 'to a large group of illustrious losers'.[50] He also took a very critical view of his own work and his publications. He considered himself to be 'too ambitious' and a researcher who 'expects perfection' from his collaborators, but this was also what he expected from himself.[51] As we have seen above, Ahlquist and Levy performed further studies on the adrenergic system to consolidate the dual adrenoreceptor concept, and his interest in receptor research and receptor blockade did not diminish.[52] But Ahlquist was extremely scrupulous and cautious in propagating his dual adrenoceptor concept, especially at an early stage of his research, as a well-tried and reliable theory. Even in 1948, he described his theory only as a 'useful' tool for further investigations:

> Although little can be said at the present time as to the fundamental nature of the adrenotropic receptor and the difference between the alpha and beta types, this concept should be useful when studying the various actions of epinephrine, the actions and interactions of the sympathomimetic agents, and the effects of sympathetic nerve stimulation.[53]

As a careful researcher, Ahlquist's reluctance to boast about a new theory was a part of his philosophy and not only a reaction to early criticism of his concept. However, in 1962, when his concept was internationally accepted (at least in basic pharmacological research), Ahlquist commented – repeating the remark in 1980 –

> One way to stimulate the search for truth is to put forth positive, dogmatic statements of belief. The reaction to these should be an attitude of complete disbelief and a quick rebuttal. If the following speculations contain grains of truth or activate others to find the truth, their objective will have been accomplished.[54]

After 1948, Ahlquist did not hesitate to confirm that his dual adrenoceptor-concept was 'an abstract concept conceived to explain observed responses of tissues produced by chemicals of various structure'. It was 'totally unnecessary from a physiological point of view'. In a strict sense, α- and β-receptors did not exist and to name them 'alpha' and 'beta' was more or less a formal decision. Instead, they would only be tools, used to help understand the functions of the human body. Furthermore, adrenergic receptors would be 'only one of the information-transferring systems'. Ahlquist viewed the rapid development of receptor subtypes with suspicion. He had debates with A.M. Lands from the department of pharmacology at the Sterling-Winthrop Research Institute in Rensselaer, New York, who, together with some colleagues, had distinguished $\beta 1$- and $\beta 2$-receptors in 1967. Then, in 1974, Salomon Langer divided $\alpha 1$- and $\alpha 2$-receptors in Buenos Aires. Ahlquist believed that 'if there are too many receptors something is obviously wrong'. In the 1970s he postulated that there would be only one adrenergic receptor. This receptor would respond to epinephrine and its α- and β-ness would be determined by the biological environment. An example for Ahlquist was the frog heart, whose receptor-types would depend on the temperature: a warm heart would have β-, a cold heart α-receptors.[55] Ahlquist told people that he had developed his concept just by chance and with an unforeseeable outcome.[56]

Ahlquist's self-critical attitude towards his receptor concept also made itself felt in his teaching of pharmacology. There are enough indications to show that teaching became more important for him than research in the late 1960s and from then on his laboratories were in a kind of disarray. It seems that he was not often in the laboratory and that he did not push research. Also, the laboratories were not up to date in the fields of molecular biology and genetic research. This changed only when Lowell Greenbaum succeeded Ahlquist in 1979.[57] The reasons for this shift of interest are not entirely clear. Perhaps it was a kind of resignation because research had not given him the rewards that he had expected, or perhaps he took this step intentionally, having done decisive work in basic research that now should be spread via medical education. What is clear is that Ahlquist is described as a devoted teacher who tried to interact with students and, above all else, cared for their needs in his later years. He tried to improve medical, and especially, pharmacological education, explaining not only his theory but also the instrumentation used and the respective experimental settings. He propagated the introduction of computers; the new building of the Medical College was planned by Ahlquist for specific educational purposes: he wanted to introduce teaching for small groups of students, with a focus on specific diseases – in the same way as medical education is carried out today.[58]

Ahlquist remained proud of his theory, and in 1962 he remarked that his students had carried the message worldwide. The main point, though, is that he did not sell his concept to the students as a decisive step in the research on the sympathetic system. In contrast, he saw it predominantly as a 'teaching aid', and there is the assumption that the concept had 'for a number of years ... primarily an educational value, simplifying the teaching of sympathetic nervous system function to medical and graduate students'. In 1973, Ahlquist even claimed that the concept was 'conceived to help teach medical students'. Indeed, at the beginning of the 1980s, Ahlquist confessed to students that he had never believed in the concept of different α- and β-receptors, but that he found it the easiest way to conceptualize the sympathetic system for medical students. When students pointed out the impact of the dual adrenoceptor-concept on the understanding of certain pharmacological processes, Ahlquist often corrected them with a phrase for which he became famous – 'It depends.' It is still an open question as to whether or not Ahlquist was reluctant to propagate his theory in the scientific community and in medical education because there had been so much criticism (as Greenbaum has suggested). In any case, the way in which Ahlquist taught his α- and β-receptors corresponded to the critical and scrupulous examination of research that had been done by others as well as by himself.[59]

Over and above the claims of rival concepts, it seems that Ahlquist's difficult character was a further hindrance in the spreading and breakthrough of his concept. He was a private man and a loner, believing in his work, but – in the words of a former student – 'not promoting himself enough'.[60]

The 'breakthrough': Sir James Black and the practical application of Ahlquist's concept

Ahlquist had begun work on the sympathetic system with the practical application of his results in mind. His early collaboration with Woodbury included clinical trials, and later in his life he maintained an interest in clinical work and kept in touch with colleagues working in medical practice. However, his receptor concept remained within the realm of theoretical pharmacological research, and for a long time there was no vision of how to apply his approach to medical practice.[61]

The therapeutic breakthrough of Ahlquist's theory and also of the receptor concept in general came with the work of Sir James Black, who, since 1957, had worked at Imperial Chemical Industries (ICI) in England. Black's research relied on an already existing cardiovascular programme in the company[62] and in this sense it was also backed by the already existing knowledge about the derivates of adrenalin. But it was also influenced by new approaches, because Black was one of the first researchers who viewed the understanding of α- and β-receptors, and the corresponding blocking substances, from the standpoint of medical practice. He was also the first to publish applicable results on this matter.[63] His story has been told many times. Below we focus on the core data of the relevant events.

Black's intention was to find an agent to combat angina pectoris, a very serious symptom of coronary heart disease, which is caused by an insufficient blood supply to the heart muscle. There were two approaches to treating this symptom: first, to increase the blood supply of the heart and, second, to reduce the cardiac oxygen consumption. Black thought about the latter possibility and developed the hypothesis that it might be possible to block the action of adrenalin on the heart. Consequently, Black performed animal experiments, measuring the heart rate as an indicator of a decrease of oxygen consumption. Black worked with isolated guinea pig hearts. In 1960, following a change in the chemical structure of DCI by one of his collaborators, John Stephenson, Black developed the substance ICI 38174, called 'pronethalol'. Black and Stephenson described their results in the *Lancet* in 1962.[64] The announcement generated clinical trials, which were promising but indicated that the substance fostered cardiac failure in some patients or worsened already existing symptoms. Although there were problems, this was the first 'β-blocker' successfully applied to patients.

Black and his team continued with their investigations using improved agents and in 1962 developed the substance ICI 45520 (propranolol).[65] In 1962, Robin Shanks, former research fellow of Ahlquist, joined Black's team. Shanks tested propranolol on animals. These trials were urgently needed as pronethalol proved to be carcinogenic in mice and was only granted a restricted licence for life-threatening conditions in the United Kingdom in 1963, eventually being withdrawn from the UK market in October 1965. Black's team, including Shanks, started with propranolol trials on man in early 1964 and published the first results in the same year. In the summer of 1964, further clinical trials were carried out, indicating that the drug was safe and effective for the treatment of angina pectoris and certain kinds of cardiac arrhythmias.[66] Propranolol, the first clinically successful β-blocker, was licensed in 1965 in the United Kingdom, only two years and eight months after its first application to an animal. This was possible on the basis of a carefully performed research plan and good collaboration between basic researchers and clinicians.[67]

Over the following years, the β-blockers were successfully improved. For example, 'cardioselective' ones were developed, inhibiting only the excitatory beta-receptors, or, in Lands's terminology, the 'β1-receptors', and not the inhibitory ones, working on the bronchial muscle or peripheral blood vessels.[68] One important early step was the increase of the therapeutic usage of the β-blockers: Black had suggested using the substance, above all, in cases of angina pectoris and acute myocardial infarction. In 1964, Brian Prichard from University College Hospital London found that pronethalol had an antihypertensive effect, and he described the therapeutic use of this substance against hypertension. Prichard quoted Ahlquist as well as Powell and Slater, Moran and Perkins and Black and Stephenson. The substance was tested on 15 patients, among them 11 with hypertension. All showed a significant fall in blood pressure, and Prichard suggested searching for a non-carcinogenic β-blocker which could be used in therapy. He and Peter M.S. Gillam proved the hypotensive effect for propranolol in the very same year. The substance was successfully applied to 24 patients, 16 of whom suffered from hypertension.[69] Since the early 1970s, together with a diuretic, β-blockers have been used widely and successfully to decrease arterial pressure.[70]

It is not the primary aim of this chapter to describe the story of the β-blockers and their usage in medical practice, with all the requisite pros and cons, but it is important to consider that acceptance of β-blockers also meant acceptance of the receptor theory. While Ahlquist could say with hindsight that Robin Shanks 'carried the idea of alpha and beta to ICI and was involved in the early studies of beta blockers in man', Black did not hesitate to point out that it was Ahlquist who gave him decisive ideas for his research on β-blockers. Black had found a chapter by Ahlquist on adrenergic pharmacology in Drill's pharmacology textbook. This chapter included Ahlquist's α- and β-receptor theory and proved to be inspirational for Black's further research.[71] In 1983 Black wrote to Ahlquist's widow that he owed 'a great deal to his [Ahlquist's] brilliant discoveries of forty years ago', that he was standing 'on the shoulders of a giant'.[72] The connections between Ahlquist's and Black's work were reflected by the prizes that both researchers won, namely the CIBA Award for Hypertensive Research and the Albert Lasker Award.[73] In 1965, according to Prichard, Ahlquist was the chairman of the first symposium that was ever held on beta-blocking agents. In 1976, when β-blockers were already fashionable in the treatment of cardiovascular diseases, the German magazine *Stern* linked the work of Ahlquist and Black. In 1988, when Sir James Black received the Nobel Prize, it was speculated that, if Ahlquist had still been alive, he would have won it jointly with Black. Ahlquist's colleague and old supporter Neil Moran fuelled this notion with a remark which he made in the same year: 'There is no question Ahlquist's concept had great impact ... I'm sure Ahlquist would have shared in the Nobel Prize if he were alive because they [Ahlquist and Black] shared the Lasker Award.' Moran also pointed out that many winners of the Lasker Award had subsequently won the Nobel Prize.[74]

Finally, it was the practical application of Ahlquist's theoretical approach by Black that popularized the receptor idea in the scientific context.[75] Increased knowledge and improved technical possibilities in pharmacology during the 1960s and 1970s have also to be considered. For example, methods were developed for fragmenting membranes mechanically with various detergents, which helped to isolate specific membrane proteins (and therewith receptors). In 1971, Lincoln Potter and Michael Raftery were able to label receptors in the torpedo fish and the eel with the help of radioactive substances (α-bungarotoxin). In 1974, radioactive β-blockers were used to label receptors in the red blood cells of frogs (Robert Lefkowitz) and of turkeys (Alexander Levitzki, Gerald Aurbach [1927–91]).[76] But these efforts were essentially supportive – it was the practical evidence in medical therapy that served to convince the scientific community of the importance of receptors. In 1948, when Ahlquist wrote his famous paper, the emphasis of research in clinical pharmacology lay on transmitter research, and this continued to be so until the late 1960s. There was only a minor shift of interest from transmitters towards receptors in these years, for example in the discussions of the British Pharmacological Society on the subject. A real change of atmosphere occurred only after the section on clinical pharmacology had been founded in 1969.[77]

Conclusions

Ahlquist developed his dual adrenoceptor theory as a child of his time. After the Second World War an optimistic mood prevailed in science, with a vision of successful research pursued with the help of new technical tools that had been developed during wartime. The supporters of the receptor concept benefited from this situation: Ahlquist's concept was a breakthrough in modern receptor research because it opened up investigations into receptor subtypes. To a certain extent, Ahlquist was the Paul Ehrlich of modern pharmacology. Like Ehrlich, he undertook many painstaking experiments to develop a new and influential receptor theory. Like Ehrlich, he was basically interested in promoting clinical therapeutics with his work, but delivered a theory which could not immediately be transferred to medical practice. Like Ehrlich, Ahlquist did not originally plan to perform research on receptors. Both men developed theories which were only based on indirect evidence of receptors. And as in the case of Ehrlich, the fate of Ahlquist's theory was influenced decisively by the context of his times and by his personality.

Furthermore, like Ehrlich's receptor theory, Ahlquist's concept was not immediately acknowledged by the scientific community. Ahlquist had problems in publishing his results, and there were difficulties in the acceptance of his theory, as it questioned the widely held belief in the cardiosympathetic drive as a stress reaction of the body (fight, flight and fright).[78] Furthermore, Ahlquist's theory contradicted other contemporary findings on amines and was both very vague and very theoretical. It was therefore understandable that it did not easily find acceptance in the scientific community of pharmacologists. Moreover, the competitive research strand on transmitters, led by Henry Hallett Dale until the late 1960s, continued to flourish.[79]

But there was also a decisive difference in comparison with Ehrlich, which aggravated Ahlquist's difficulties. It was his personality, his frank and direct comments on scientific matters, as well as his pessimistic and sceptical attitude, which hampered Ahlquist's own work. Whereas Ehrlich did not hesitate to illustrate his invisible 'pluralism' of receptors, Ahlquist was self-critical and in no way the type of man to promote himself successfully. He needed the help of others, for example Neil Moran, and it was only the practical application of his dual adrenalin receptor concept by Sir James Black that brought his theory into the mainstream of the scientific community of pharmacologists.

Whether Ahlquist truly did not receive enough recognition for his work remains debatable, especially as the appreciation of scientific discoveries is linked with contemporary notions and evaluations. But his case does demonstrate that science is not an isolated area, and that its performance can be fully understood only when considering its historical context, the social position of the researchers involved, and the rules of 'science in action'. It is quite often the case that scientists do not get appropriate rewards for their work, or that they lose out against apparently less talented colleagues. The decision-making of editorial boards of journals and of committees of scientific institutions is often guided not only by the quality of the research concerned but also by considerations of its political and strategic relevance and its relationship to dominant scientific theories or paradigms. This is still true in our time.

7
The Emergence of Molecular Pharmacology

> Although some progress has been made in the isolation of receptors and their characterization in chemical and physicochemical terms, for the most part receptors must be regarded as hypothetical entities, even though the receptor concept lies at the heart of pharmacology.
>
> (Bowman and Rand, 1980: p. 39.15)

Isolation and identification of Langley's 'receptive substances' (Ehrlich's 'side-chains') required the effort of many research groups with expertise in diverse areas of the biomedical sciences, including pharmacology, biochemistry, physiology and molecular biology. Serendipity also facilitated the purification of receptors and their subsequent biochemical and molecular characterization.

This chapter will consider some of the figures involved in the research and the technical developments from the late 1950s to the early 1990s leading to the *cloning* of receptor proteins. We will focus on the nicotinic acetylcholine receptor, because this is the 'receptive substance' that Langley had studied so extensively in the autonomic nervous system and at the neuromuscular junction, and because it illustrates the principles by which other receptors were isolated. The enormous increase in the number of scientists and the volume of information generated on receptors during this period also necessitates that we concentrate on the acetylcholine receptor to illustrate the emerging complexities of these neurotransmitter gated proteins. Finally, we will touch upon the implications of these recent observations for the concept of the receptor in pharmacology and for the development of new therapeutics.

Location of receptors

The work of Langley and Clark, in particular, had suggested that receptors were located only in certain specialized regions of tissues. The nicotinic 'receptive substances', for example, were found at autonomic ganglia and the neuromuscular junction. In 1926 R.P. Cook observed that the dye, methylene blue, blocked the effects of acetylcholine on the frog heart (an atropine-like action) before it also stained the heart muscle. This antagonist action was reversed by washing the dye from the bath in a heart that remained deeply stained blue. Re-applying methylene blue to stained hearts again inhibited the action of acetylcholine. These results suggested that methylene blue combined reversibly with receptors at the cell surface.[1] Technically more sophisticated experiments by J. Del Castillo and Bernard Katz in the 1950s, using electrophysiological techniques, showed that depolarization of frog muscle only occurred when they applied acetylcholine locally to the end-plate region (where the motor nerve innervates the muscle) and that microiontophoresis of acetylcholine *into* the cell had no effect.[2] These observations and subsequent data confirmed the idea that transmitter receptors were located at the surface of cell membranes and most especially at synapses. These *physiological* receptors showed a very high degree of selectivity for endogenous biomolecules (transmitters or hormones) and exogenous drugs, and mediated highly specific cellular (pharmacological) responses. Such *physiological* receptors were found to be the most abundant in the body and the targets of many drugs.

Isolation of brain synaptosomes

Biochemical, biophysical and microscopic techniques provided considerable data showing that cell membranes were composed of a lipid bi-layer[3] and integral proteins that were fluid within the lipid membrane. This so-called 'fluid mosaic model' of the cell membrane was proposed by S.J. Singer and J.L. Nicolson in 1972. The proteins of cell membranes constitute carrier molecules, enzymes and receptor proteins. Isolation of receptors required the isolation of cell membranes and their integral proteins.

Fundamental and elegant experiments by Eduardo De Robertis (1913–88), the director of the Institute of Cell Biology at the University of Buenos Aires, were conducted over two decades to visualize and isolate brain synapses. De Robertis trained in medicine and in 1939 was awarded the University Gold Medal for achieving the highest grades in his class at the University of Buenos Aires. Following graduation, he continued his training in the USA (at Johns Hopkins, the University of Chicago and Massachusetts Institute of Technology) before returning to his home country in 1957 'after a long exile' to become director of the Institute of Cell Biology. His task was to establish a neuroscience research group in Argentina.[4]

From 1954 onwards, De Robertis and colleagues used the electron microscope to make visible the complex ultrastructure of the synapse and, subsequently, developed the biochemical methods to isolate the nerve endings. The isolated nerve endings, later named 'synaptosomes' by V.P. Whittaker and colleagues (1964), were shown to be formed from the pre-synaptic nerve terminal and the post-synaptic cell membranes containing the *chemosensitive* (receptor) sites.[5]

Separating synaptosomes from other brain constituents into a homogeneous fraction involved several physical and chemical steps including homogenization of the cerebral cortex, hyposmotic shock to rupture cell membranes, exposure to detergents to dissolve the lipid membrane and high speed ($\geq$100,000 times gravity) centrifugation through a gradient of sucrose. Observation of the '*dissected*' synaptic region using the electron microscope revealed the presence of synaptic vesicles and the sub-synaptic web (the post-synaptic density thought to contain the synaptic receptors), but confirmation that receptors remained present was problematic. In fact, unlike the biochemical methods used to separate an enzyme, where its presence could be bioassayed *in vitro* at several stages of cell fractionation, the response mediated by receptors (for example, contraction in a muscle) was lost when the cell was broken down. The presence of receptors was therefore not certain in synaptosomes.

To indicate the presence of receptors, several groups therefore used *radioactively-labelled* cholinergic drugs, such as dimethyl-^{14}C-d-tubocurarine, [^{3}H]atropine, [^{3}H]pilocarpine to 'label' nicotinic and muscarinic receptors in the mammalian brain nerve endings.[6] These experiments clearly showed radioactivity present in the synaptosomes and especially in the fraction containing the sub-synaptic membrane. Separation of the membrane protein from the membrane lipid by chromatography showed radioactivity enriched in the proteolipid components[7] suggesting the chemoreceptor substances were probably protein.[8] This critical body of work also demonstrated that the synaptic terminal could be dissected free of other parts of nervous tissue and that receptors could be isolated, opening up the way to their purification.[9] However, the identity of the synaptosomal components labelled by the cholinergic drugs used was not known since they were non-selective agents and bound to the receptor, to the cholinesterase enzymes that break down acetylcholine, and to other cell constituents.

Snake venoms and fish electric organs: isolation of the cholinergic receptor protein

Nature and serendipity greatly facilitated isolation and purification of the nicotinic receptor protein by providing investigators with two very important tools: one, an animal source highly enriched with the nicotinic acetylcholine receptor and two, a potent snake toxin that binds selectively and essentially irreversibly to the receptor protein.

The electric organs found in rays such as *Torpedo californica* and *Torpedo marmorata* and in the electric eel, *Electrophorus electricus*, are formed from modified skeletal muscle, known as *electroplax* and enable these fish to generate large currents to stun and kill prey. These electric organs were used therapeutically for pain control in ancient Greece and Rome and also studied by the pioneers of bioelectricity and electrophysiology, Alessandro Volta (1745–1827), Luigi Galvani (1737–98) and Emil Du Bois-Reymond (1818–96). The cholinergic nature of transmission at the electromotor synapse of *Torpedo marmorata* was established at Arcachon, France, in 1939 by Feldberg, Fessard and Nachmansohn soon after transmission at the neuromuscular junction had been shown to be cholinergic.[10] In the *Electrophorus* electric organ, there are 1–10 billion identical acetylcholine-containing synapses; in *Torpedo*, 400 g of fresh electric tissue can yield milligram quantities of receptor protein.[11] The electroplax cells are depolarized by cholinergic agonists (that is, drugs that activate the receptor) through increased membrane permeability to sodium, potassium and calcium ions, and these responses are blocked by cholinergic antagonists, like tubocurare (arrow poison). The electroplax is 500–1000 times richer in cholinergic synapses than muscle and was therefore a highly enriched model for the study of nicotinic acetylcholine receptors.[12]

C. Chagas (1910–2000), the founder of the Institute of Biophysics at the Federal University of Rio de Janeiro, Brazil, and colleagues (1958) made the first attempts to isolate the cholinergic receptor from the electric organ of *Electrophorus* but were unsuccessful. By using very high concentrations of radio-labelled gallamine, a cholinergic antagonist drug, these researchers precipitated undefined polysaccharides (complex carbohydrates) which are abundant in electric fish. Isolation of the receptor, therefore, needed a drug that would bind selectively and irreversibly.

Within five years, Chen Yuan Lee (1915–2001) and colleagues, working at the National University of Taiwan, had discovered that the venom from cobras and the *Elapidae* snake, *Bungarus multicinctus*, contained a component that blocked neurotransmission at the neuromuscular junction and caused paralysis in animals bitten by these snakes.[13] Snake bites were a major health problem in Taiwan. Lee and co-workers at the Institute of Pharmacology conducted internationally recognized studies on the isolation, composition and pharmacological properties of poisonous snakes found in the farming areas of Taiwan.[14] The purified polypeptide from *Bungarus multicinctus*, called α-bungarotoxin, was shown to bind to the end-plate region of the neuromuscular junction with very high affinity and to irreversibly block the nicotinic acetylcholine receptor but not to bind to the cholinesterase enzyme.[15] It was also observed that d-tubocurarine protected the receptor against the α-bungarotoxin blockade. From these findings, Lee and co-workers concluded that α-bungarotoxin combines selectively and irreversibly with the cholinergic receptor.[16]

Interestingly, during the last decade of his life, Lee also became actively involved in political reform in Taiwan and was unanimously elected to be the first chairman of the new Taiwan Independence Party in 1996. When the entrenched Kuomintang government was defeated by the Democratic Progressive Party in 2001, Professor Lee was appointed senior adviser to the newly elected president. For his scientific and social achievements Chen Yuan Lee is still held in high regard in his native Taiwan as well as internationally.

Molecular properties of the nicotinic receptor: 1970–90

Two key neuroscientists, in quick succession, then utilized the potent and highly selective α-bungarotoxin to isolate the acetylcholine receptor substance from the electric organ. In June 1970, the French neuroscientist Jean-Pierre Changeux, working at the Pasteur Institute in Paris, reported the isolation of the acetylcholine receptor protein from the electric organ of *Electrophorus electricus*.[17] Eight months later, in February 1971, Ricardo Miledi, then professor of biophysics, and colleagues at University College London reported the first isolation of the acetylcholine receptor protein of *Torpedo* electric tissue in the journal *Nature*.[18] Notably, De Robertis and co-workers first reported isolation of cholinergic receptor proteolipid from the electric tissue of *Torpedo* and *E. electricus*[19] at a meeting in Sweden in February 1970. However, they had used less selective and low affinity radio-labelled cholinergic ligands, such as hexamethonium, and critically had not demonstrated that this 'receptor material' was specifically from the electroplax membranes. This work was therefore quickly superseded by the studies of Changeux and Miledi and their respective colleagues.

Changeux's and Miledi's contributions to the understanding of the role of receptors and ion channels in synaptic transmission in the nervous system have been both fundamental and complementary. Changeux was educated at the Ecole Normale Supérieure and received his doctorat d'etat de sciences naturelles in Paris in 1964. Changeux's interests (developed as a graduate student in the laboratory of the 1965 Nobel Laureate for Medicine or Physiology, Jacques Monod, 1910–76) were in the concept of allosteric regulation of enzymes and receptors. Amongst many original observations since 1970, Changeux and his colleagues reported extensively on the biochemical properties of the nicotinic receptor protein, including the amino acids that contribute to the acetylcholine binding site and the ion channel pore; on visualization of the purified receptor protein with the electron microscope first reported as a 90 A (9 nanometre) rosette with hydrophilic centre;[20] and on evidence that the receptor was composed of 5 protein subunits.[21] Some of this work will be described below. Changeux has received many awards for his outstanding contributions to science including the Balzan Prize, Linus Pauling Medal, Max Delbruck Medal, Louis Jeantet Prize, Gold Medal of the CNRS, Richard Lounsbery Prize and the Gairdner Foundation International Award.

Ricardo Miledi graduated in medicine from Universidad Nacional Autonoma de Mexico (1955). He then received research training as a Grass Scholar at the Marine Biological Laboratories in Woods Hole, Massachusetts, USA, and as a Rockefeller Foundation Fellow with Sir John Eccles (1903–97), the 1963 Nobel Laureate for Physiology or Medicine, at the John Curtin School of Medical Research in Canberra, Australia. From 1958 until his appointment as Distinguished Professor at University of California Irvine in 1985, he was a faculty member at University College London, where he headed the department of biophysics. During a long and fruitful collaboration with Sir Bernard Katz (1911–2003), the Noble Laureate for Medicine or Physiology in 1970, he produced an outstanding series of papers analysing both pre-synaptic and post-synaptic mechanisms of transmission, especially at cholinergic synapses. This work included proof of the quantal hypothesis of synaptic transmission;[22] demonstrating the role of voltage-dependent calcium entry in the pre-synaptic terminal;[23] the fine localization of neurotransmitter receptors in the post-synaptic membranes of muscles and the squid giant synapse; and the use of noise analysis to prove that synaptic potentials are due to the opening of discrete ion channels.[24] Recognition of Miledi's scientific contributions have also been marked by the awarding of numerous honours and prizes, including the King Faisal Foundation International Prize for Science, the Principe de Asturias Prize for Scientific Research (Spain's most prestigious award in the sciences), and the Royal Society Royal Medal (also called the Queen's Medal). In fact, Miledi is, to date, the only Mexican-born scientist to be a member of the prestigious Royal Society of London.

Receptor purification, biochemical and biophysical properties

The two papers published by Miledi and colleagues and Changeux and co-workers marked the start of an enormous increase in the number of scientists working on receptors, the volume of work published on this concept, and an increased competitiveness in receptor research.

During the 1970s several groups invented or developed the methods to isolate and purify cholinergic receptors, especially from the fish electroplax, and determined important biochemical and physical properties of the receptor. Essentially, purification of the nicotinic receptor involved homogenizing and 'dissolving' the cell membranes in mild detergents (such as Triton X 100 and deoxycholate) or organic solvents such as chloroform. The crude extracts were then incubated with radio-labelled α-bungarotoxin (either triturated or iodinated) and the molecular components separated on the basis of size through a sucrose gradient, or the material was passed through a chromatography column and eluted using a variety of solvents to separate different fractions.

A modification to the latter method by Richard Olsen, then an NIH post-doctoral fellow, and Jean-Claude Meunier, working in Changeux's lab, was to link a cholinergic ligand, CT 5263, to the filter beads to create an 'affinity chromatography' column. The column could then selectively bind the cholinergic receptor as it passed over the beads.[25] The eluted material was further divided into its various components by electrophoresis by which proteins were separated on a gel according to their size and electrical properties. By sampling many different fractions from the column and incubating them with α-neurotoxin, the cholinergic receptor protein was purified several thousand fold. These methods resulted in the isolation of highly purified extracts of receptor material and became the routine methods to yield toxin-tagged receptor material.

An elegant use of extracted receptor material, by Patrick and Lindstrom (1973) later showed that immunization of rabbits with purified *E. electricus* nicotinic acetylcholine receptor led to the production of antibodies to the receptor and flaccid paralysis in the animals. These data clearly demonstrated that the nicotinic acetylcholine receptor isolated from the fish electric organ was the physiological receptor at the mammalian neuromuscular junction and raised the hypothesis that *myasthenia gravis* could be an autoimmune disease – an idea now firmly established by multiple observations.

By the start of the 1970s there was general agreement that the cholinergic receptor was a protein complex and a distinct entity from acetylcholine esterase.[26] In contrast, considerable debate occurred during the decade over the size and the number of protein *subunits* making up the acetylcholine binding site and the ion channel. For example, initial studies by Miledi and co-workers[27] suggested the *Torpedo* cholinergic receptor was perhaps greater than 200,000 daltons and composed of several (perhaps 2 or 4) protein subunits each of molecular weight of 80,000. In contrast, Meunier, Olsen and colleagues (1972) in Changeux's lab calculated a molecular weight of 540,000 for the *Electrophorus* receptor on the basis of its separation through a chromatography column, but only about half this weight when estimated from sedimentation rate in a sucrose gradient. The smallest labelled subunit, corrected for binding of toxin, reported by these authors was estimated to be 48,000 daltons. The discrepancies between the Miledi and Meunier estimates might have occurred because they were using different species. However the two different estimates of the receptor size in the Meunier study were not easily resolved and over the next few years several other groups reported a broad range of molecular weights for the receptor complex.[28]

A seminal paper by Hucho and Changeux in 1973, however, reported their isolation of the receptor using an affinity column and gel electrophoresis together with a method to cross-link the receptor subunits. From their data, these authors (correctly) proposed that the *E. electricus* receptor was a protein assembly of *five* subunits associated together to form a *polymer* with a total molecular weight of 275,000. Moreover, these authors also proposed that two different polypeptide chains combined to form the polymer (in a ratio of 3:2 subunits). By 1978, Reynolds and Karlin had used a rigorous method and confirmed a molecular weight of 250,000 daltons and from their evidence and that of previous work also correctly proposed that four (not just two) different polypeptide chains, termed α, β, γ and δ made up the receptor complex in electroplax of *T. californica*. The receptor had two ligand (acetylcholine) binding sites involving the α-subunits.

During this period of intense and productive biochemical characterization, studies by Katz and Miledi using electrophysiological techniques also revealed several fundamental biophysical properties of the acetylcholine receptor. Katz had been awarded the Nobel Prize in Physiology or Medicine in 1970 for his 'discoveries relating to chemical transmission of nerve impulses' and in particular for showing that the neurotransmitter acetylcholine was released from the pre-synaptic terminal in packets or 'quanta'.

Katz and Miledi now addressed the mechanism by which activation of the cholinergic receptor led to depolarization (excitation) of the post-synaptic cell. Using intracellular microelectrodes, these two leading neuroscientists recorded changes in membrane potential (membrane noise) that resulted from the application of acetylcholine to the endplate region of the frog somatic neuromuscular junction – the prototype chemical synapse. Hypothesizing a steady-state depolarization of the endplate from an iontophoretic application of acetylcholine, Katz and Miledi proposed that the membrane noise (that is, the tiny fluctuations in the depolarization) resulted from the statistical 'bombardment' of single acetylcholine receptors. From this 'noise analysis' method, Katz and Miledi then estimated the elementary voltage change resulting from activation of a single acetylcholine receptor channel to be 0.29 μV.[29] Subsequently, they also estimated the average opening time of the channel to be around 1 millisecond and the single channel conductance to be around 100 Pico Siemens.[30] This data finally settled the long-running debate that the receptor was a channel pore, not a carrier.[31]

Development and application of the patch-clamp recording technique (1976) by Erwin Neher and Bert Sakmann a few years later, in which the discrete single ion channel openings and closings could be visualized (for the first time) from membrane patches, was remarkably consistent with the estimations by Katz and Miledi. Moreover, these two German cell physiologists were awarded the 1991 Nobel Prize for Physiology or Medicine for conclusively establishing with their technique that ion channels do exist and for revolutionizing the study of ion channels.

Cloning of the nicotinic acetylcholine receptor

By the beginning of the 1980s, it was established that the nicotinic acetylcholine receptor (found in the electroplax and at the neuromuscular junction) was a pentameric protein complex forming a ligand (acetylcholine) binding site and ion channel through which cations (sodium, potassium and calcium) move across the cell membrane. Four different polypeptides termed α, β, γ and δ subunits of molecular weights 40,000, 50,000, 60,000 and 65,000 daltons, respectively and in the ratio of $2\alpha : 1\beta : 1\gamma : 1\delta$ formed the receptor-channel complex. From electron microscopic studies, in particular from Changeux's lab, these subunits were apparently arranged in a pseudo-symmetrical 'rosette' fashion around a central pore and with an overall diameter of 9 nm.[32] Reconstitution of these protein subunits into artificial lipid membranes also displayed all of the pharmacological and physiological properties of the native cholinergic receptor.[33] Sufficient biochemical and biophysical data were thus available to enable the isolation of the genes encoding for nicotinic acetylcholine receptor subunits and the age of molecular pharmacology emerged into the ascendancy.

A critically important publication in 1980 from M.A. Raftery's laboratory at the California Institute of Technology in Pasadena reported on the micro-sequencing of the polypeptides comprising the acetylcholine receptor in *T. californica*.[34] These authors identified the first 54 to 56 amino acids in each of four polypeptides purified from the electroplax receptor. The results showed that the receptor proteins were highly similar in composition (*homologous*) but also clearly confirmed that four distinct proteins formed the receptor complex as had been thought on the basis of their separation on gels and columns. More importantly, these results provided the hitherto cryptic information that would lead quickly to the cloning of the nicotinic acetylcholine receptor subunits.

Between October 1982 and April 1983, during an especially fruitful time in the development of the concept of receptors, Shosaku Numa (1929–92) and colleagues in Japan published three articles in the journal *Nature* reporting the identification of four genes encoding the α, β, γ and δ subunits of the electroplax nicotinic acetylcholine receptor of *T. californica*. The 54 to 56 amino acid long sequences identified by Raftery's study for each of the receptor subunits were sufficient for Numa and co-workers to construct short synthetic sections of the genes (so-called degenerate oligonucleotide sequences) that encode just a small part of each polypeptide subunit. By labelling these oligonucleotide probes with radioactive phosphate (P32) and incubating them with a library of approximately 200,000 genes compiled from the electroplax genome, these authors identified four distinct but related genes. These four genes were sequenced, and the full length proteins deduced from this genetic code. The similarity in the nucleotide bases and their position in the genes (sequence homology) indicated that they had descended from a single common ancestor by gene duplication.[35] In 1985 Numa and colleagues then also utilized the frog egg (*Xenopus laevis* oocyte) to express the four cloned nicotinic receptor subunits and show electrophysiological responses upon their activation by acetylcholine, confirming a fully functional receptor ion channel.[36]

The competition to identify the genetic code for the nicotinic receptor proteins was clearly significant: over the same short period that Numa's papers were published, S. Heinemann and his group at the Salk Institute in California also published two studies in the *Proceedings of the National Academy of Sciences*, in July 1982[37] and February 1983[38] in which they sequenced a clone encoding the γ subunit of the acetylcholine receptor of *T. californica*. These studies overlapped in substance and in time, so that neither group cited the work of the other in these high profile journal publications. It is also possible that neither group was aware of the other's work.

Implications of the isolation and cloning of nicotinic receptors

By the end of the 1980s, several important conclusions had emerged from the cloning 'revolution' and from electron microscopic images of isolated nicotinic receptor crystals produced by N. Unwin and colleagues at Stanford University Medical School.[39] First, a model of the 3-dimensional structure of the nicotinic receptor was deduced showing that '5 subunits are arranged symmetrically around the ion channel having their axes approximately perpendicular to the membrane plane'.[40] Viewed from the 'top' of the receptor, the subunits were arranged clockwise α, β, α, γ and δ. The overall dimensions of a nicotinic receptor were around 9 nm in diameter and 14 nm long, with receptors traversing the cell membrane and clustered in a lattice formation at the post-synaptic cell density.

A second consequence of the cloning of the nicotinic receptor subunits, and later the cloning of other neurotransmitter receptors such as those for GABA-A receptors by E. Barnard and colleagues at the University of Cambridge, UK, and P.H. Seeburg at the University of Heidelberg in Germany,[41] was the realization that these receptors shared common ancestral genes and therefore belonged to a receptor *superfamily*. Indeed, the combination of structural and functional information also provided a more sophisticated system for the classification of numerous, apparently diverse, receptors into a small number of major receptor superfamilies. Perhaps even more unexpectedly, the molecular data showed that many different genes encoded for receptor subunits. This led to the realization that hundreds, if not thousands, of receptor subtypes could exist in the body. In addition, data emerging by the end of the 1980s revealed that mutations in genes coding for receptors could lead to dysfunctional receptors and ion channels. Such mutations in physiological receptors and ion channels were termed *channelopathies*. Evidence accumulated over the last decade indicating that human neurological diseases, such as epilepsy, are strongly associated with channelopathies in GABA receptor subunits and that congenital myasthenia is associated with mutations in nerve and muscle nicotinic receptor protein subunits.[42]

Third and most pragmatically, the possibility that many hundreds of receptor subtypes existed in the human body with distinctive functions energized the pharmaceutical industry at the start of the 1990s to search for a new array of drugs to target these sites. Intriguingly, however, as molecular pharmacology has risen, the number of new drugs approved over the past decade has greatly decreased, whilst the cost of their development has doubled. An additional and related consequence of the new science of molecular pharmacology has been the emergence of *pharmacogenomics* – that is the customization of drugs to an individual's genetic makeup. The expectation with pharmacogenomic analysis is to improve the therapeutic response of patients, especially in complex disorders such as schizophrenia (in which improvement is highly variable and unpredictable) whilst at the same time minimizing the adverse or unwanted side-effects of drugs.[43] However, these developments are still too recent to further consider their historical significance and impact at this time.

Less than 100 years after Langley had postulated the 'receptive substance', the nicotinic acetylcholine receptor had been isolated, purified, identified, sequenced, and cloned. Research during the past 25 years has furthermore led to the recognition of a plethora of receptor families, of receptor subtypes, and even 'orphan' receptors. Increasingly, receptors turned from hypothetical entities into objects of material reality, which were targeted as defined sites of pharmacological intervention. Ehrlich's 'magic bullet' has now multiplied to hundreds and potentially thousands of new therapeutic agents.

Conclusions

In our introduction, we discussed the role of constructivist theory for the history of scientific ideas in general and particularly for the history of the receptor concept. But did we follow a track which enabled us sufficiently to explain the historical construction of the receptor concept? Receptors, as the last chapter abundantly shows, play a major role in modern biomedicine, especially in the field of pharmacology. If we are living in a golden age of positivist modern science, then perhaps the indications of future successes can be identified in the history of the receptor concept. To quote a contemporary immunologist: 'The side-chain theory ... was the distant forerunner of clonal selection.'[1] Perhaps so, but the future does not determine the past, and in the following sections we will show that the theoretical framework presented at the beginning of our book is indeed useful for obtaining a deeper understanding of the history of receptors.

Medicine *in* culture: receptor concepts and cultural contexts

Undoubtedly, receptor research over the last 30 to 40 years has contributed greatly to the diagnosis and treatment of human diseases and thereby to the effectiveness of modern medicine. The theory of receptors, as it stands today, fits very well with the specific context of current biomedical knowledge. But all results of scientific work depend on their historical context and their cultural setting. Many adherents of early twentieth-century scientific medicine, for example, thought the receptor concept irrelevant to medicine, because they did not believe in the importance of chemistry for diagnostic and therapeutic procedures. We do not know to what extent future generations of physicians and biomedical researchers will continue to accept our present concept of receptors.

What is seen as a success in our times might come to be seen as a wrong path in a few decades.

In retrospect, we can identify three important periods in the history of the receptor concept: first the last decades of the 'long nineteenth century' (from the construction of the 'receptors' in 1878 until 1918), second the interwar period and the Second World War (1918–45), and third the post-war period and the establishment of medical standards that have since shaped our experience of biomedicine (1945–70). In the following paragraphs, we will briefly summarize these periods.

From about 1878 until the end of the First World War, there was the period of the concept's introduction. Discussions on the topic mainly focused on the work of Paul Ehrlich and John Newport Langley who first developed and later defended their theories. Ehrlich lived until 1915, Langley until 1925, and during their lifetimes they promoted their theories and thereby kept them alive. But they were not pharmacologists – Ehrlich came from the bacteriological research tradition of Robert Koch's group in Berlin, and Langley was a student of the physiologist Michael Foster at Cambridge University. Their 'receptors' and 'receptive substances' originated from different cultural contexts. Inspired by Koch's fight against microbes, Ehrlich's 'receptors' were part of an immunological microcosm of defensive bodily substances. Langley's 'receptive substances' were tools for understanding the nervous system of animals and human beings. The two theories were linked in so far as both Ehrlich and Langley saw their 'receptors'/'receptive substances' as part of a fundamental explanation of the physiology of the animal and human body. Because of its different disciplinary origins and – so to speak – the 'split invention', the receptor concept had an uncertain future. Its success in pharmacology was anything but guaranteed.

Second, we recognize a period of debate, of criticism and of defence of the concept, which lasted from around 1918 to 1945. Critics and their alternative ideas gained credibility during this period because Langley and Ehrlich could only provide indirect evidence for their theories. Their techniques did not permit any direct demonstration of the existence of receptors. However, it was not only the supporters of the receptor concept who had a credibility problem, but also other scientists – among them the critics of the receptor concept – who constructed and attempted to prove alternative theories (chemical or physical) explaining drug binding to organs and cells.

In this context the 'physical theory' of drug action appeared to be the most plausible solution. It rested on the nineteenth-century research tradition of experimental pharmacology and was supported internationally by a majority of pharmacologists. The 'physical theory' had its basis in experimental physiology – a well-established medical discipline of the nineteenth century – not in chemistry, where findings from the chemical industry (from about 1850) were only slowly incorporated into academic medicine. The experimental setting of physiology seemed to be more suited to a scientific pharmacology which aimed for therapeutic impact. The development of a strong research strand around transmitter substances underlined this quest for therapeutic success. For example, transmitter research during the interwar period led to drugs for the treatment of disabling vegetative symptoms – like vomiting or high blood pressure.

Overlapping groups of physical theorists and transmitter researchers established a network that opposed those who wanted to invade pharmacology with ideas of chemical binding to cell receptors. This is all the more remarkable inasmuch as the transmitter researchers themselves defended, in the 'soup versus sparks' debates, claims about a *chemical* nature of neurotransmission against those who adhered to the *physical* concept of electrical transmission from cell to cell.[2] The receptor concept only survived because these networkers in pharmacology were unable to give a fully satisfactory alternative explanation for the highly selective binding of drugs and dyestuffs to cells and tissues and for their very specific biological effects. Discussions about different theories without any definite answers created an atmosphere of unease among pharmacologists. Torald Hermann Sollmann and Paul John Hanzlik, and above all Alfred Joseph Clark and his followers, benefited from this situation. Clark's contribution was especially important in promoting the quantification of drug actions on cells. His mathematical approach and the creation and analysis of concentration-effect curves, first comprehensively published in 1933, led to a significant development of the receptor theory, defended at the time by a small but prestigious group of pharmacologists. Specific ideas of the quantitative approach, such as receptor occupancy, competitive and non-competitive antagonism, and the concept of partial agonists, were successively introduced and accepted – at least in academic pharmacology.

Third was a period of a slow acceptance of the receptor concept in pharmacology, lasting from approximately 1945 to 1970. The eventual breakthrough came later, because traditional pharmacologists still dominated the field. Even such an innovative mind as Raymond P. Ahlquist, whose research could rely on better facilities than had been available to Langley or Ehrlich, struggled with the old and still influential research strands in pharmacology. Although Ahlquist opened the door to a more effective usage of the receptor concept, when differentiating adrenergic receptor subtypes, a lack of credibility remained. This credibility was only achieved when evidence could be produced – and agreed upon – that those entities called 'receptors' were molecular units of the cell. But from the 1950s onwards technical innovations proved crucial. Receptors were successively isolated, purified, identified as glycoproteins, sequenced and cloned. At the same time – and importantly for the eventual breakthrough of the concept – receptor-subtype specific drugs were introduced into clinical medicine.

With hindsight, we might detect a gradual acceptance of the concept, yet even in the third period of receptor history divergent attitudes prevailed. For most of its history, the receptor idea was condemned as much as it was admired. From a historical perspective, it was an old idea, with roots reaching back into the mechanistic thinking of seventeenth-century physiologists. It was seen either as a theoretical speculation or as a product of 'hard' science based on meticulous research. The different points of view depended – and depend – on the more general attitudes of scientists or on the orientation of the scientific school to which they were (are) committed. For example, the attitude of Walther Straub, who argued above all against Ehrlich's receptor concept, shows him as a proponent of the nineteenth-century pharmacological school of Buchheim and Schmiedeberg, which was based methodologically on contemporary experimental physiology.

It is notable in this context that it is not possible to assign the various key arguments in the receptor debate to specific historical periods. Ehrlich and Langley had supporters from the beginning, when they started to publish their ideas about side-chains and receptors. And even today there are critical voices emphasizing the model character of the various receptor theories.

From our historical analysis, the factors that determined the fate of the receptor concept appear to have been complex. We do not know what would have happened if Ehrlich had kept his position as Frerich's senior physician at the Charité Hospital in Berlin, or if Langley had decided to stick to other physiological research problems after 1900. From the perspective of the history and philosophy of science we can say in general terms that the breakthrough of specific ideas or concepts is a question of probability, dependent on contemporary views, on the distribution of influence and power and, not least, on luck. This certainly applies to the history of the receptor concept. Ehrlich's and Langley's theories were resisted by the scientific community of pharmacologists, and the adherents of the receptor theory were treated as outsiders in scientific medicine; it would have been hard to predict that such a concept had any future, harder still to predict its role in the development of effective therapeutics later in the twentieth century. When Alfred Joseph Clark and some followers revived the concept in the 1930s and 1940s, they also experienced resistance; their theories were regarded as the products of personal predilections for studying problems of basic science – almost a luxury at a time when it seemed necessary to promote clinical pharmacology as a provider of effective new therapies. After the Second World War, it needed the stamina of John Henry Gaddum to defend the receptor concept against dominant researchers in the field, such as the Nobel Prizewinner Henry Hallett Dale, for whom receptors were pointless speculations in comparison to the fruitful research on transmitters. It was hard for the supporters of the chemical receptor to integrate their concept into a pharmacological theory of drug binding which also considered the physical or dynamic explanations of this process. Finally, it seems, the practical application of the receptor concept in pharmacotherapy created an atmosphere of trust and credibility; Sir James Black's work on receptor-blocking agents, especially on the β-blockers, was particularly important.

What does this history teach us with regard to the nature and practice of modern science and medicine? This book is not the first historical analysis of cultural influences on scientific research, but it deals in depth with an idea that has become highly fashionable and that is the basis of our current understanding of many biological processes at the cellular and sub-cellular level. Moreover, *cultural context* is still an important part of an ongoing debate. When confronted with the social or cultural 'construction' of science in a historical context, researchers in science and medicine often feel affronted.[3] It seems to them as if their scientific work is being degraded to a plaything of power relations in society. Historians of science and medicine – in the view of practising physicians and scientists – do not understand the methodology of hard science and how to achieve 'objective' knowledge in the laboratory. The consequence of accepting cultural constructivism, some believe, would be to hinder or even prevent scientific research. This is a misunderstanding, however, of the aims of science studies and especially of the constructivist approach. Science and scientific medicine remain effective systems for ordering the world, but consideration of the limitations of research leads to a more critical view of scientific findings and consequently a more responsible use of scientific knowledge. For example, medical practitioners can better evaluate the presentation of statistics by pharmaceutical companies when they know the historical case studies elucidating the various influences on the production of seemingly 'objective' scientific knowledge. We would argue that a critical attitude in medicine is strengthened through critical analysis provided by the history of science and medicine. The history of the receptor concept demonstrates this point. Even a well-established and apparently self-evident theory may need review and critical analysis to avoid stagnation.

Medicine *as* culture: medical theory and clinical practice

This book is also concerned with a basic problem of modern medicine, discussed among physicians and scientists from the very beginnings of scientific medicine in the nineteenth century: the relationship between medical practice and medical theory, or between clinical medicine and laboratory medicine. Both parts of scientific medicine developed rapidly during the nineteenth century, and we need to go back into the history of this period to explain the impact of the problem on the history of the receptor concept. Here, we are dealing not only with the various cultural influences on the receptor concept but also with the impact of the receptor concept *on* medical culture (medicine *as* culture).

The basic features of clinical practice were developed in the so-called Paris School of Medicine between approximately 1790 and 1830.[4] This school had a pragmatic approach to working at the sickbeds (of poor, hospitalized patients) and identified traditional medical theories with a medical elite that had supported the *Ancien Régime*. Auscultation, that is, the investigation of the patient with the stethoscope, and percussion, the tapping of the chest or the abdomen in order to detect irregularities, represented this new medicine, as did post-mortem examinations to identify pathological changes in the organs. Success in treating patients rested very much on the skills and experience of the physician. For some physicians, medicine was an art, which could be practised only by a talented individual who was able to see, feel, smell and – sometimes – even taste the disease.

Laboratory medicine developed in the middle of the nineteenth century. Experimental physiology and pathological anatomy, in particular, became leading disciplines of modern medicine. The main idea was that scientific medicine would be based upon data collected empirically in the laboratory. The lab was the place where elements of nature could be isolated and investigated under conditions controlled by the experimenter. It enabled the study of the laws of nature by removing the distracting influences that misled observers outside the laboratory. Medicine, according to proponents of laboratory science such as Rudolf Virchow, had to rest on solid knowledge about the basic morphological conditions and physiological laws of the human body. Knowledge about these matters was *in this context* regarded as necessary in order to treat patients in a rational and successful way.[5]

However, the two branches of scientific medicine – clinical practice and laboratory work – followed different principles, and there was a serious problem of communication between them as well as institutional alienation at the end of the nineteenth century in Germany, and also more generally in other Western countries.[6] Among historians of medicine, and many physicians and scientists, there has been a tendency to see this development as a persistent feature of modern medicine, one that still has an impact today. The idea is that since about 1850 the medical practitioner has been using a different language from that of the laboratory researcher (for example, the molecular biologist), and that two different types of medicine with different therapeutic concepts have run in parallel. Furthermore, the introduction of modern techniques and machinery associated with the medical laboratory into clinical medicine has apparently alienated the patient from the physician.[7] Historians of science have envisaged laboratory work as performed by Virchow or Pasteur as efforts to build artificial zones for reconstructing the appearance of diseases and immunological defence mechanisms in experimental animals.[8] However, these artificial lab environments did give impulses to medical practice and this historical view has tended to overlook the connections and the growing interdependency between the laboratory and the clinic from the last decade of the nineteenth century. Medicine came to rely increasingly on 'experts', who acquired specialist knowledge in a particular field of medicine, with the aim of promoting diagnostics and therapy through their (laboratory) findings.[9] There are several examples of this new type of researcher who appeared from about 1880, including the chemist Fritz Haber (1868–1934).[10]

These experts often occupied a grey area between the laboratory and the clinic and/or had close links with clinicians who performed laboratory research themselves. Also, these experts were sometimes able to coordinate the work of different partners in the health-care system, namely politicians, drug companies, general practitioners and last but not least university institutes. Paul Ehrlich as a 'would-be-clinician' and laboratory researcher belonged to this group of experts. From the start of his career, Ehrlich's ambition was to influence the general course of medicine – initially as a clinician with laboratory experience, and later as a laboratory researcher who worked on human biology – to develop a new kind of therapy, armed with 'magic bullets'. Ehrlich brought together the work of politicians, such as his sponsor Friedrich Althoff, pharmaceutical companies like the *Farbwerke Hoechst*, and the capacities of the

Institute for Infectious Diseases of his academic sponsor, Robert Koch.[11] In a narrower sense, John Newport Langley can also be counted as one of these experts. As student and successor of the Cambridge physiologist Michael Foster, Langley followed, to a degree, his teacher's vision of developing a new biology of man when he tried to uncover the secrets of the 'autonomic nervous system'. Like Ehrlich, Langley looked beyond his own discipline and presented his receptor concept to pharmacologists both nationally and internationally.

In conclusion, the history of the receptor concept illustrates the way in which the boundary between laboratory medicine and clinical practice became increasingly blurred after 1900. The early history of the receptor concept reflects the beginnings of 'biomedicine', which had its full breakthrough after the Second World War. The idea of applying biological data and laws derived in the laboratory to everyday therapy posed a challenge for clinicians and laboratory researchers, and its development helped in the construction and acceptance of the receptor concept. And the receptor concept, in turn, in the long run promoted interdisciplinary work and new institutional settings. Multiple disciplines of biomedicine and biology have incorporated receptor research since the 1960s and – to a certain extent – collaborated in the investigation and treatment of specific human diseases. The receptor concept has become so strong that it is now difficult to think of modern medicine without it. In analogy to recent approaches in the science studies the receptor concept may be described as a sort of 'boundary idea',[12] which was useful for industrial innovation, for medical therapy and for the achievement of political aims in science. Therefore, the history of the receptor illustrates important characteristics in the evolution of biomedicine, a development which could only take place because of symbiotic connections between theoretical and practical fields in medicine. The receptors shaped medical culture in the twentieth century.

Taking all this into account, the history of the receptor concept provides a case study of the impact of cultural factors on the emergence and breakthrough of a medical innovation or scientific idea, and it stands for the development of a specific medical culture in medicine, namely the development of modern laboratory research combined with investigations in clinical medicine. The history of the receptor concept *in culture* and the receptor concept *as* a specific medical *culture*[13] tells us a lot about modern medicine in an age of transformation from the morphological to the physiological, when insights into disease conditions were accomplished by the recognition of disease processes.

Notes

Introduction

1. For general literature on the history of the receptor concept see the notes below and the first chapter of this book.
2. See Bowman (1999).
3. Drews (2002).
4. 'World Health Organization Model List of Essential Medicines', 15th edition (2007), http://www.who.int/medicines/publications/essentialmedicines/en/; NCPA report, Bartlett (2000), http://www.ncpa.org/oped/bartlett/sep0400. html; for US figures, see Hagist and Kotlikoff (2007), www.ncpa.org/pub/st/st286.
5. Recent examples include Schüttler (2003) and Mould (1993).
6. See the example of the scientific genetics community in the United States and Germany (Harwood 1993). For a more recent example see Robert E. Kohler's description and analysis of the community of geneticists working on the fruit-fly drosophila ('fly group') (Kohler 1994; 1999).
7. Stanton (2002, esp. pp. vii–x) and Löwy (1993a, pp. v–viii). See also Pickstone (1992).
8. Hagner (2001, esp. p. 23).
9. Daniel (2001).
10. Morrell (1993, esp. p. 124).
11. Morrell (1993; 1972).
12. Moraw (1988, p. 3).
13. For details see the list of archival sources at the end of the book.
14. Concerning oral history and its problems, see Benison (1971), Perks (1990), Thompson (1991; 2000), Thompson and Perks (1993), Ward (1995), Perks and Thompson (1997) and Singer (1997).
15. Ibid.
16. Stannard (1961) and Leake (1975, p. 59).
17. Harig (1974) and Debru (1997).
18. Galen, *De simplicium medicamentorum temperamentis ac facultatibus* 7.12 and 11.1. Thanks to Philip van der Eijk (Newcastle) for these references.
19. Stein (1997) and Watson (1966).
20. Galen, *De naturalibus facultatibus* 1.13. Thanks to John M. Forrester (Edinburgh) for pointing out this passage. Compare Galen, transl. Brock (1928, pp. 49–71).
21. Keyser (1997).
22. Stille (1994, pp. 13–17, 374–6), Leake (1975, pp. 27, 97–9), and Riddle (1992).
23. Stille (1994, pp. 71–89, 382).
24. Maehle (1987, p. 109) and Houlston (1784). Thanks to Stan Frost (Durham) for this latter reference.
25. Maehle (1999, pp. 231–3).
26. Maehle (1999, pp. 135–8, 145–6).

27. Dehmel (1996).
28. Earles (1961), Maehle (1987; 1999).
29. Bynum and Porter (1988).
30. Wittern (1991).
31. Stille (1994, pp. 383–6) and Knight (2003).
32. Bickel (2000, pp. 37–51).
33. Grmek (1973).
34. Bickel (2000, pp. 94–121) and Kuschinski (1968).
35. Compare Schmiedeberg (1867, p. 274) and Stille (1994, pp. 373–4).
36. Bynum (1970).
37. Parascandola (1974) and Chapter 3 below.
38. See Prüll (2003a).
39. Stille (1994, pp. 387–8) and Chapter 3 below.
40. Parascandola (1986), Parascandola and Jasensky (1974), Robinson (2001, pp. 143–70, 199–218) and Silverstein (2002). See also Cozzens (1989).

1 Paul Ehrlich and his Receptor Concept

1. For current views on the history of the receptor concept see Silverstein (2002), Parascandola and Jasensky (1974), Parascandola (1986), Bennett (2000), Maehle et al. (2002), Prüll et al. (2003) and Maehle (2004a; 2004b). An earlier version of this chapter has been published by Prüll (2003b).
2. For the constructivist view on the history of science see Golinski (1998) and Lenoir (1997).
3. This chapter makes use of the printed works of Ehrlich between 1878 and 1905, which can be found in Himmelweit (1956; 1957; 1960). It relies also on parts of the unprinted obituary of Ehrlich, his laboratory books, his laboratory notes and his correspondence. The latter materials are kept in the Rockefeller Archive Center (RAC) in New York. The chapter also utilizes materials from the Archive of the Humboldt-University in Berlin and from the State Archive of Prussian Cultural Heritage (*Geheimes Staatsarchiv Preußischer Kulturbesitz*: GStA PK), Berlin. For details see the list of archival sources at the end of the book.
4. Malkin (1993, pp. 129–37) and Clark and Kasten (1983).
5. Bäumler (1997, pp. 24–30, 31–6), Jokl (1954, esp. p. 972) and Goldsmith (1934, esp. p. 69). Concerning Ehrlich's staining work on 'plasma cells' and 'mast cells' see also Ehrlich (1877; 1878), Dolman (1981, p. 296) and Silverstein (2002, p. 2).
6. Prüll (2003a, pp. 204–65) and Maulitz (1978).
7. These cells originate, as we know today, from so-called B-lymphocytes and produce antibodies.
8. 'Mast cells' are cells in the connective tissues, containing among others the antibody immunoglobuline E, which plays an important part in allergic reactions.
9. Concerning Ehrlich and chemistry, see also Dale (1956, esp. p. 2) and Goldsmith (1934, pp. 69–70).
10. '...eine bestimmte chemische Beschaffenheit der Zelle selbst ...', in Ehrlich (1878). See the quotation on p. 40; p. 75 (English translation).

11. Ehrlich (1885).
12. Ehrlich (1897a). For a contemporary description of Ehrlich's side-chain theory, see Aschoff (1902, esp. pp. 1–25). See also Heymann (1928) and Silverstein (1989, esp. pp. 64–6, 94–9).
13. Ehrlich and Morgenroth (1900b).
14. Parascandola (1981, esp. p. 28).
15. Ehrlich (1902, esp. p. 595; English translation, p. 618).
16. Parascandola (1986, pp. 134–41).
17. See Bynum (1970).
18. It is not possible in this chapter to give a full account of the history of the application of the receptor concept to drug binding and pharmacology. See above all Parascandola (1981, pp. 21, 30–3, 35).
19. The best general account of Ehrlich's life is Dolman (1981). Dolman also gives an overview of the literature on Ehrlich up to 1980. The writing on Ehrlich has remained largely hagiographic. See for example, Bäumler (1997). There are several reasons for this. As Ehrlich was Jewish, all public written testimonies of his life were erased by the Nazi government after 1933; and his estate has only comparatively recently become accessible for research. Furthermore, hagiographic accounts were presumably promoted by the new scientific optimism of the post-war decades, for example Witebsky (1954). For the history of research on Ehrlich, see also Dale (1949, pp. xiii–xx; 1956) and Bäumler (1997, pp. 5–9). The serious recent historiography of Ehrlich largely comprises papers on special aspects of his work.
20. See for example Koch (1924) and Dale (1949, p. xvi).
21. See for example Jokl (1954, p. 974) and Klose (1954, esp. p. 425).
22. Wassermann (1914).
23. '... zur Erkennung der wissenschaftlichen Entwicklung Ehrlichs sehr wertvoll'; see Michaelis (1919, esp. p. 165, also pp. 167–8).
24. Eckmann (1959) and Marshall (1995, esp. p. 565).
25. Travis (1989), Lenoir (1988), Silverstein (1999, 2002).
26. Parascandola and Jasensky (1974) and Parascandola (1986). Also concentrating mainly on Ehrlich's ideas see Moulin (1991, esp. pp. 74–97).
27. Dale (1956, p. 9).
28. See Sauerteig (1996).
29. Franken (1994) and Classen et al. (1995).
30. Travis (1989, p. 393). For the Medical Clinic of the Charité in Berlin, see Winau (1987, pp. 139–42, 198–9).
31. Dolman (1981, p. 296) and Ehrlich (1883d).
32. Ehrlich (1882a).
33. Franken (1994, pp. 59–62).
34. Ehrlich (1883c; 1883a; 1883b; 1886a; 1900a; 1901a).
35. See for example a case study on phosphor poisoning: Ehrlich (1882b).
36. Ehrlich (1882c).
37. See as an overview Ehrlich and Lazarus (1900).
38. Ehrlich (1882d; 1882e). For Robert Koch's work on tuberculosis and tuberculin, see Gradmann (2001, 2004).
39. Ehrlich (1886b).
40. Ehrlich (1885, esp. p. 415).
41. Ibid., pp. 368, 419, 422, 430.

42. Ehrlich (1891b, esp. p. 166).
43. Note, in Acta der Friedrich-Wilhelms-Universität Berlin. Habilitationen von 1880–89, Medizinische Fakultät – Dekanat –, No.1342/1, Archive of the Humboldt-University, p. 203.
44. Ehrlich and Lazarus (1900, p. 213).
45. Ibid., p. 258.
46. Bäumler (1997, pp. 60, 68). See Margaret Goldsmith's remark on Ehrlich's life after his appointment at Frerich's clinic: 'The next seven or eight years were one of the most fruitful and satisfying periods of Ehrlich's life' (Goldsmith 1934, p. 71).
47. Ibid., p. 72. The view expressed in the *British Medical Journal* concerning Ehrlich's work at Frerich's clinic, that practical clinical work did not suit Ehrlich, is one of the rare comments on this topic in the literature. See Obituary (1915a, p. 349). For the time with Frerichs, see Dolman (1981, pp. 296–7).
48. Winau (1987, pp. 198–200).
49. Ibid., p. 200.
50. Goldsmith (1934, pp. 73–4) and Dolman (1981, p. 297). For Carl Gerhardt, see furthermore Professor Carl Gerhardt (1902), p. 1, in: Acta betr. die Anstellung des Geheimen Medicinal Raths und Professors Dr. Gerhardt als dirigirender Arzt und Director der 2. medicinischen Universitätsklinik, 1885, Kgl. Charité-Direction, No. 437, Archive of the Humboldt-University, p. 37.
51. Marquardt (1949, pp. 27–8) and Bäumler (1997, pp. 68–9).
52. Liebenau (1990, esp. p. 66).
53. Ehrlich (1891c; 1891d; 1892). See also Lazarus (1922, pp. 34–5).
54. Goldsmith (1934, p. 76).
55. Dolman (1981, p. 297) and Ehrlich and Guttmann (1891b, pp. 7–12).
56. Ehrlich and Leppmann (1890) and Bäumler (1997, p. 76). On the further impact of Ehrlich's research on the therapy of psychiatric disorders with methylene blue, see Healy (2002, pp. 44–7).
57. Ehrlich and Guttmann (1891a).
58. Weindling (1992, esp. pp. 170–2, 175–8, and see the quotation on p. 170). For Robert Koch and his research programme, see also Gradmann (2005).
59. See the descriptions of the clinical trials, in Ehrlich et al. (1894, esp. pp. 57–60) and Ehrlich and Kossel (1894).
60. Silverstein (2002, p. 42). Although Silverstein's biography on Ehrlich's receptor concept is almost exclusively devoted to a history of ideas, it gives occasional hints on the importance of the social setting for the development of Ehrlich's career and the receptor concept.
61. Bäumler (1997, pp. 90–3) and Dolman (1981, p. 297). For Ehrlich and his early involvement in the development of serum therapy against diphtheria, see Ehrlich et al. (1894), Ehrlich and Kossel (1894), Ehrlich and Wassermann (1894) and Ehrlich (1894).
62. Goldsmith (1934, p. 76). In the 1892 report on the work of Koch's institute, Ehrlich is mentioned only once as a voluntary assistant. See Ueber den Bericht ... (1892) [off-print from the Deutsche Medizinische Wochenschrift, No. 4-7 (1892)], in Acta betr. die Einrichtung und die Verwaltung

des (staatlichen) Institutes für Infektionskrankheiten in Berlin, vom Januar 1892 bis Dezember 1898, in GStA PK I HA Rep. 76 Kultusministerium, VIII B, No. 2893, pp. 63–75, esp. p. 71. In the records of Koch's institute, kept by the Berlin Charité-Hospital for the years 1893 to 1895, Ehrlich's name does not appear at all. See Acta betr. das Institut für Infectionskrankheiten, Kgl. Charité-Direction, No. 2205, 1893–1895, Archive of the Humboldt-University.

63. The Prussian Minister of Science and Education to Robert Koch, 9 February 1895; *Bericht über die Thätigkeit des Kgl. Instituts für Serumforschung und Serumprüfung zu Steglitz*. Juni 1896–September 1899. Zur Einweihung des Königl. Instituts für experimentelle Therapie Frankfurt/M., Jena, Fischer, 1899, in Acta betr. das Institut für experimentelle Therapie zu Frankfurt a.M., vom Februar 1895 bis Dezember 1900, GStA PK I HA Rep.76 Kultusministerium, Vc Sekt.1, Tit. XI, Teil II, No.18, vol. 1, pp. 1, 189–203, here 203.

64. For Friedrich Althoff, see vom Brocke (1991; 1980, pp. 9–118).

65. Paul Ehrlich to Friedrich Althoff, 27 July 1907, in Nachlass Althoff B, GStA PK VI HA, Rep.92, No.33, pp. 217–22, esp. p. 219. See furthermore Eckart (1991, on Ehrlich esp. pp. 398–401).

66. See for example Hedwig Ehrlich to Friedrich Althoff, 13 September 1903, in Nachlass Althoff B, GStA PK VI HA, Rep.92, No.33, pp. 71–4; also Goldsmith (1934, p. 74). It is not the place here to discuss the difficulties between Ehrlich and Behring. This is described and analysed in the secondary literature on Ehrlich. See furthermore Linton (2005). Although Linton's biography of Behring is quite hagiographic, he does not deny Behring's contributions to the difficulties with Ehrlich.

67. Goldsmith (1934, pp. 77–8). Concerning the organization of serum testing and Ehrlich's collaboration with representatives of medicine, politics and industry, see Hardy (2006).

68. Ehrlich (1890; 1891a).

69. Ehrlich (1897a). The citations in the notes below refer to the German original. For the impact of this paper on Britain, see also Plimmer (1897).

70. Ehrlich (1897a, pp. 89–93).

71. Ibid., pp. 93, 96. See also Lazarus (1922, p. 36).

72. Ehrlich (1897a, pp. 93–106) and Heymann (1928, p. 1258). The idea of cellular regeneration stemmed from Ehrlich's cousin Carl Weigert and was developed in contact with the latter (see Heymann 1928). For the toxin-antitoxin reaction and Ehrlich's work, see also Mazumdar (1974).

73. Ehrlich (1897a, p. 94). For the history of the 'key-lock' metaphor in molecular biology and enzymes, see Cramer (1997) and Kay (1993, pp. 112–13, 165–6). See furthermore Farber (1981) and Remane (1984).

74. For this period of Ehrlich's work, see also Dolman (1981, p. 298).

75. For the work done in the *Institut für Serumforschung und Serumprüfung* in Steglitz, see Dönitz (1899). Dönitz remarked that the biological application of the chemical theory of side-chains was the basic idea of all research activities of the institute.

76. Goldsmith (1934, p. 78).

77. For the life of Morgenroth see Bäumler (1997, p. 329) and Dönitz (1899, p. 360).

78. Prüll (2003a, pp. 229–33).

79. 'Bitte mir auch die pyrodin-thiere zeigen!', 'Wo bleibt der Affe!', see Note, 6 March 1900, p. 13; Note 1900 (no day and month), p. 30, in Zettel Buch I, 1900, February 1 to 1900, December 26, folder 1,2,3, box 7, series II, (3) – 1,2,4 (No. of transcripts & index copy), Paul Ehrlich Collection, Rockefeller University Archives, RAC.

80. Ehrlich (1898).

81. Ehrlich and Morgenroth (1899a).

82. '... ständig wechselnden Chemismus', see Ehrlich and Morgenroth (1899b, esp. p. 162; English translation, esp. p. 170). See also Prüll (2003b, esp. pp. 344–5).

83. 'Wir wollen im Folgenden stets, um eine grössere Kürze des Ausdrucks zu ermöglichen, diejenige bindende Gruppe im Protoplasmamolekül, an welche eine fremde, neu eingeführte Gruppe angreift, allgemein als "Receptor" bezeichnen', see Ehrlich and Morgenroth (1900a, esp. p. 196; the quotation is taken from the English translation, p. 205).

84. Ibid.; Ehrlich and Morgenroth (1900c).

85. See the quotation in Silverstein (2002, p. 80). The introduction of the term 'receptor' in 1900 is underestimated by Silverstein, who describes it merely as a 'reminder that the side-chain theory holds that antibodies are cell receptors' (ibid., p. 105).

86. Von Dungern (1914).

87. 'Eigenthümlichkeiten des Receptorenapparates', see Ehrlich and Morgenroth (1901a, esp. p. 238; English translation, p. 249).

88. Ibid., p. 244; Ehrlich and Morgenroth (1901b, esp. p. 272).

89. Lazarus (1922, pp. 41–2).

90. Ehrlich and Morgenroth (1901b, pp. 256, 258).

91. Ibid., pp. 268, 273.

92. Ehrlich and Sachs (1905, esp. p. 434).

93. Ehrlich (1904, esp. p. 317).

94. Ehrlich (1901b, esp. p. 312).

95. Ibid., p. 313.

96. Ehrlich had the idea of applying a specific bacterial serum to different animal species with a different receptor apparatus. Therewith he aimed to produce antibodies against all bacterial receptors of this microbe. He thought that in doing this he could obtain a map of all receptors of this microbe. He would be able then to combat it effectively with a therapeutic serum produced from the mixture of the produced antibodies. See Ehrlich and Morgenroth (1901b, p. 259).

97. '... eine neue bedeutungsvolle Richtung der biologischen Forschung eröffnet', see Ehrlich (1904, p. 320). See also Dolman (1981, pp. 298–9).

98. Dönitz (1899, pp. 376–84).

99. Ehrlich (1903a, esp. p. 383).

100. McCormmach (1982, esp. pp. 20–1) and Chalmers (1986, pp. 110–11, 189).

101. '... die Probe des Versuchs aufs Beste bestanden'. See Ehrlich (1901b, p. 306).

102. Ehrlich and Morgenroth (1901a, pp. 234–5).

103. Ehrlich to Veit, 22 January 1903, in Direktor, Ehrlich X, 1902, December 23 to 1903, January 26, folder 1,2,3,4,5,6, box 22, series V, (3-complete) – 1,2,4 (2-incomplete) (No. of transcripts & index copy), Paul Ehrlich Collection, Rockefeller University Archives, RAC, pp. 169–71, esp. p. 171.

104. Ehrlich and Morgenroth (1901b).
105. Weindling (1992, pp. 172–4, 176 (see the quotation here), 178–82).
106. See Ehrlich's remark on Roux and Büchner, which seems to be the beginning of the discussions about the side-chain theory: Ehrlich (1897b).
107. Ehrlich and Morgenroth (1900c, pp. 214, 218, 220–3). For Bordet see Bordet (1899a; 1899b; 1900). See also Aschoff (1902, pp. 96–8), Silverstein (1989, pp. 99–104) and the detailed analysis of Eileen Crist and Alfred I. Tauber (1997, esp. p. 329); also Prüll (2003b, p. 347).
108. Laborzettel, 2 × 1901, in Zettel Buch II und Carcinom, 1900, December 25 to 1901, September 21, folder 1,2,3, box 8, series II, (3) – 1,2,4 (No. of transcripts & index copy), Paul Ehrlich Collection, Rockefeller University Archives, RAC, pp. 292/3. For some general remarks on the dispute between Ehrlich and Bordet see also Silverstein (2000, esp. pp. 37–9). See also the comments of August von Wassermann (1914, pp. 148–50).
109. Der 'wirkliche spiritus rector'; mit seiner 'hinreissenden persönlichkeit'. Es sei 'jammerschade, dass ein solcher experimentator und klarer kopf wie Roux so in den mysticismus und den russischen nebel gerathen musste!' See Ehrlich to Salomonsen, Kopenhagen, 24 February 1899, in Copir Buch III. Ehrlich, 1899, February 21 to 1899, July 31, folder 1,2,3,4, box 5, series I, (3) – 1,2,4 (No. of transcripts & index copy), Paul Ehrlich Collection, Rockefeller University Archives, RAC, pp. 13–16. Metchnikoff's theory of phagocytosis (digestion) of bacilli through leucocytes and the latter's ability to produce antitoxin was seen as attack on the 'side-chain theory' by Ehrlich; see Aschoff (1902, pp. 28, 42–3, 103–4, 115–17). For some general remarks on the dispute between Ehrlich and Metchnikoff see also Silverstein (2000, pp. 35–7) and Günther (1954, pp. 68–107, esp. pp. 75–81).
110. '... in der ganzen Linie sich anerkennung verschafft haben', see Ehrlich to Kobert, Rostock, 1 December 1902; Ehrlich to Eulenburg, Berlin, 25 November 1902, in Direktor, Ehrlich IX, 1902, September 25 to 1902, December 23, folder 1, 2, 3, 4, 5, 6, box 21, series V, (3-complete) – 1, 2, 4 (2-incomplete) (No. of transcripts & index copy), Paul Ehrlich Collection, Rockefeller University Archives, RAC, pp. 434–7, 376–7, quotation on p. 377.
111. 'Hocherfreulich war es mir, Sie wieder als einen so warmen freund der theorie zu erkennen, noch mehr aber, dass Sie durch diesselbe zu so neuen und fundamentalen gesichtspunkten gekommen sind'; 'dass jeder Unbefangene, der die Literatur liest, Sie zu den absoluten Gegnern zählen muss'; see Ehrlich to Welch, Baltimore, 20 October 1902; Ehrlich to Fuld, Halle, 27 October 1902, in Direktor, Ehrlich IX, pp. 117–18, 182–7, esp. pp. 182–3.
112. Ehrlich to Arrhenius, 1 January 1903, in Direktor, Ehrlich X, 1902, December 23 to 1903, January 26, folder 1, 2, 3, 4, 5, 6, box 22, series V, (3-complete) – 1, 2, 4 (incomplete) (No. of transcripts & index copy), Paul Ehrlich Collection, Rockefeller University Archives, RAC, pp. 71–9. Arrhenius and Madsen (1903); Arrhenius (1915, esp. pp. 110–39). For a detailed description of the Ehrlich-Arrhenius controversy see Rubin (1980).
113. For example, Ehrlich to Römer, 8 July 1903, in Direktor, Ehrlich XI, 1903, March 13 to 1903, July 17, box 23, series V (mostly typescripts, no transcripts) (No. of Transcripts & index copy), Paul Ehrlich Collection, Rockefeller University Archives, RAC, pp. 445–8; Ehrlich to S. J.

Meltzer, 30 December 1903, in Direktor, Ehrlich XIII, 1903, December 21 to 1904, June 13, box 23, series V (mostly typescripts, no transcripts) (No. of Transcripts & index copy), Paul Ehrlich Collection, Rockefeller University Archives, RAC, pp. 44–6. See also further letters of Ehrlich in this file.

114. Ehrlich to Friedrich Althoff, 12 September, presumably 1904, in Nach-lass Althoff A I, GStA PK VI HA, Rep. 92, No. 258, pp. 15–16, and esp. pp. 22–30. For the controversy between Ehrlich and Arrhenius/Madsen, see also Mazumdar (1995, pp. 202–13). On Arrhenius's 'Ionization Theory' as an example of his physical chemistry strand, see Malkin (2003, pp. 337–44, esp. pp. 338–40).

115. Dolman (1968, esp. pp. 80–1), Gruber (1901a, pp. 1827–30, 1880–4; 1901b, pp. 1924–7, 1965–8; 1903a, p. 2297; 1903b, pp. 791–3) and Gruber and von Pirquet (1903, esp. p. 1194).

116. Ehrlich to Römer, 8 July 1903, in Direktor, Ehrlich XI, 1903, March 13 to 1903, July 17, box 23, series V (mostly typescripts, no transcripts) (No. of Transcripts & index copy), Paul Ehrlich Collection, Rockefeller University Archives, RAC, pp. 445–8; '... ich mache ihm nur zum Vorwurf, dass er bei seinem Theoretisieren zuviel Phantasie und zu wenig Kritik walten lasse'. See Gruber (1903c, esp. p. 1825).

117. Michael Hubenstorf (2001, p. 140) described Gruber as a 'gifted polemicist' (*begabter Polemiker*).

118. 'Nur die begleitenden Umstände der Lebensvorgänge sind unserer Forschung zugänglich'. See Gruber (1903c, p. 1827).

119. Ehrlich (1903a; 1903b). For the debate with Gruber see Silverstein (1989, pp. 104–7) and Mazumdar (1995, pp. 123–35).

120. See the detailed analysis in ibid., pp. 107–278, esp. pp. 136, 216–17, 147, 226.

121. Ibid., p. 255.

122. See Weindling (1992), Gradmann (2000), Schmiedebach (1999) and Münch and Biel (1998).

123. Marquardt (1924, pp. 29, 36, 40–2, 54–5).

124. Cambrosio et al. (1993, esp. pp. 667–9, 684) and Crist and Tauber (1997, pp. 336–7, 346–53).

125. Marquardt (1924, pp. 28–9) and Dolman (1968, p. 81).

126. For Doyle's work and its interpretation see Laura Otis's brilliant analysis (Otis 2000, esp. pp. 91–8).

127. Cambrosio et al. (1993, p. 699).

128. Schlich (2000).

129. Ehrlich (1900b, p. 187).

130. Ehrlich to Althoff, August 1903, in Nachlass Althoff B, GStA PK VI HA, Rep. 92, No. 33, pp. 142–3, esp. p. 142.

131. Crist and Tauber (1997, p. 333). Crist and Tauber also point out that Ehrlich never abandoned his side-chain theory (p. 325).

132. For Ehrlich's extremely sensitive reaction to critics see the more or less hagiographic account of Walter Greiling (1954, esp. pp. 120–2). Research on chemotherapy was, above all, carried out in the Georg-Speyer House, newly erected in 1906. This institute was attached to Ehrlich's institute. See Das Speyer-Haus in Frankfurt a.M. (Berliner Klinische Wochenschrift, 1906), in Nachlass Althoff A I, GStA PK VI HA, Rep.92, No.258, p. 66.

133. Liebenau mentions that the institute in Frankfurt gave Ehrlich the opportunity to conduct 'controlled bedside trials'. Besides the problem of transferring modern terms and methods back to the past, one has to consider that Ehrlich himself was no longer attached to a clinical unit and that he relied on clinical colleagues to perform therapeutic human experimentation. See Liebenau (1990, p. 70).

134. '... nachdem ich so lange serumtherapeutische fragen fast ausschliesslich bearbeitet habe, mich wieder etwas mehr meinem alten lieblingsgebiet theorie der histologischen und biologischen färbung zuzuwenden.'; 'Nach dem langen immunitäts zauber komme ich jetzt wieder dazu, mich meinem alten lieblingsgebiet der farbstoffe wieder etwas zuzuwenden...', see Ehrlich to the Badische Anilin- und Sodafabrik, Ludwigshafen, 15 November 1898; Ehrlich to Nietzki, 24 December 1898, in: Copir Buch II. Direktor, 1898, November 11 to 1899, February 21, folder 1, 2, 3, 4, box 4, series I (3) – 1, 2, 4 (No. of transcripts & index copy), Paul Ehrlich Collection, Rockefeller University Archives, RAC, p.p. 69, 242.

135. See Himmelweit (1956; 1957; 1960).

136. See especially Ehrlich's letters, in Copir Buch II. Direktor, 1898, November 11 to 1899, February 21, folder 1, 2, 3, 4, box 4, series I (3) – 1, 2, 4 (No. of transcripts & index copy), Paul Ehrlich Collection, Rockefeller University Archives, RAC.

137. '... zunächst bei kopfschmerzen, vagen rheumatoiden seuchen, gonorrhoe und cystitis'. See Ehrlich to Albert Neisser, no date, in ibid., pp. 93–4.

138. 'Schliesslich ist es doch auch wohl nicht gar zu schwer – bei Eurer grossen übung in diesen dingen – die ungefähre dosis bene tolerata heraus zu bekommen'. See Ehrlich to Neisser, 30 November 1898, in ibid., p. 128.

139. Ehrlich to Neisser, 9 December 1898; 7 January 1899 (ibid., pp. 187–80, and esp. pp. 180, 299). See furthermore Elkeles (1985, pp. 135–48).

140. '... alle diese herren treiben die farbentherapie eigentlich mehr aus gefälligkeit mir gegenüber, als aus innerer überzeugung.' See Ehrlich to Iwanoff, Petrowsk, 30 June 1899, in Copir Buch III. Ehrlich, 1899, February 21 to 1899, July 31, folder 1, 2, 3, 4, box 5, series I (3) – 1, 2, 4 (No. of transcripts & index copy), Paul Ehrlich Collection, Rockefeller University Archives, RAC, pp. 398, 399–402, 403, esp. pp. 398–9.

141. Carl Weigert to Ehrlich, 27 March 1897, in Nachlass Althoff A I, GStA PK VI HA, Rep. 92, No. 258, pp. 15–16, esp. p. 15. As a compensation for the contributions of the city of Frankfurt to the erection of Ehrlich's institute, Ehrlich had to examine body fluids and organ specimens from patients of the City Hospital. See Bäumler (1997, pp. 113–14); Ehrlich (1907) in Nachlass Althoff B, GStA PK VI HA, Rep. 92, No. 33, pp. 244–55, here pp. 251–3. Ehrlich's cousin Carl Weigert influenced the administration of the city of Frankfurt in favour of Ehrlich's and Althoff's plans.

142. Ehrlich to Althoff, 23 February 1900, in Nachlass Althoff B, No. 33, pp. 114–15.

143. 'Da ich selbst nicht in der Lage bin, derartige Untersuchungen an einem grösseren Krankenmaterial durchzuführen, habe ich es für meine Pflicht gehalten, die Gesichtspunkte klarzulegen und so die Basis für die Bearbeitung eines Gebietes zu schaffen, dessen Bedeutung für die Pathologie

und Therapie vielleicht erst nach Jahren voll gewürdigt werden wird.' See Ehrlich (1901b, p. 315).

144. Ehrlich to Althoff, 1 January 1905, in Nachlass Althoff B, GStA PK VI HA, Rep. 92, No. 33, pp. 150–6, here p. 153 (emphasis in the original).
145. Joannovics (1915, esp. p. 940).
146. Münch and Biel (1998, pp. 13–14).
147. Parascandola (1981, esp. p. 30). See also Ehrlich's papers on chemotherapy, especially the Harben Lectures of 1907 (Himmelweit 1960).
148. Jokl (1954, pp. 972–4).
149. Prüll (2003a, pp. 230–1).
150. Szöllösi-Janze (1998, see the introduction on pp. 9–22; 2000) and Gradmann (1998).
151. Parascandola and Jasensky (1974) and Parascandola (1986).
152. Parascandola (1974, esp. pp. 55–6).
153. This was already registered by contemporaries, see, for example, Obituary (1915b, esp. p. 525). It is considered also in the secondary literature on the receptors: see, for example, Rubin (1980, pp. 400–7) and Cambrosio et al. (1993, pp. 662–99).
154. Ehrlich saw the side-chain theory as a 'unifying bond' (*einigendes Band*) of the work of all departments in the Frankfurt institute. See Ehrlich (1907, p. 251). See furthermore 'Paper on the Cancer Research of Ehrlich', anonymous, Frankfurt, 7 April 1906, in Nachlass Althoff A I, GStA PK VI HA, Rep. 92, No. 258, pp. 32–8, here esp. pp. 37–8; Lenoir (1988, pp. 79–82).
155. See Ehrlich and Sachs (1909; 1910) and Ehrlich to Althoff, 18 March, presumably 1907, in Nachlass Althoff B, GStA PK VI HA, Rep. 92, No. 33, p. 329. Although the basic idea of chemical specificity survived in immunology after the death of the master, this was not true in general for his whole receptor concept. The latter remained controversial, as Ehrlich's matured theory could not be used in total to explain the action of the living organism, see Mazumdar (1974, pp. 18–21); also Bulloch (1960, pp. 275–9).
156. See also Wassermann (1909, esp. p. 247).

2   The Development of the Concept of Drug Receptors in the Physiological Research of John Newport Langley

1. On Langley's contribution to the development of the receptor theory, see also Parascandola and Jasensky (1974) and Parascandola (1986). An earlier version of this chapter was published by Maehle (2004b).
2. See French (1975) and Rupke (1990).
3. On the research of this group see Geison (1978).
4. See Langley (1907a, pp. 236–40), Du Bois-Reymond (1926), Fletcher (1926; 1927) and Geison (1973; 1978). For the contents of the practical physiology course, see also Foster and Langley (1899).
5. Langley (1875b, p. 404). See also Langley (1875a). The paralysing action of curare on nerve endings had been suggested by Claude Bernard. See Olmsted (1939, pp. 223–8).
6. Langley (1875c, quotes from pp. 194–5). See also Geison (1978, pp. 242–3).

7. Langley (1876, p. 180).
8. Luchsinger (1877, p. 488).
9. Ibid., pp. 491–2.
10. Langley (1878, p. 367).
11. Rossbach (1879, pp. 33–8).
12. Langley (1880, p. 19).
13. For a summary of his findings in this area see Langley (1898).
14. Foster was secretary of the Royal Society from 1881 to 1903, served on various Royal Commissions, and represented the University of London in Parliament from 1900 to 1905 (initially as a Liberal Unionist, then as a Liberal). He acted practically as scientific adviser to the government. Foster published no research papers after 1876. See Langley (1907a, pp. 244–5), Thiselton-Dyer (1907) and Geison (1978, p. 7). In 1883, Foster had been elected professor of physiology at Cambridge. He was the only candidate. See Election of Professorships under Statute B. Chapter IX, Statutes (1926), Chapter XIV, UA CUR O.XIV.54, Dept. of Manuscripts and University Archives, Cambridge.
15. See Klein et al. (1883), Langley (1883a; 1883b), Langley and Sherrington (1884) and Langley and Grünbaum (1890). For the scientific background see Clarke and Jacyna (1987).
16. Langley (1915), Ackerknecht (1974) and Geison (1978, pp. 313–19).
17. Langley and Dickinson (1889).
18. See, for example, Langley (1899; 1903a; 1903b; 1905a).
19. Election of Professorships under Statute B. Chapter IX, Statutes (1926), Chapter XIV, UA CUR O.XIV.54, Dept. of Manuscripts and University Archives, Cambridge; Readers and Lecturers from 1878 to 1929, vol. V, Physiology – Talmudic, UA CUR 113.5, Dept. of Manuscripts and University Archives, Cambridge, pp. 22, 25, 32.
20. Ibid., pp. 17, 31. Box II: History of Lab Building 1912–14, Letter Accounts, to Prof. Langley, Deed of J. Physiol. from Mrs. Langley. Papers in connection with building of Physiol. Lab, Dept. of Physiology, University of Cambridge.
21. Langley and Dickinson (1890a).
22. Langley and Dickinson (1890b). On the tradition of physiological experimentation with alkaloids, going back to the work of François Magendie in the early nineteenth century, see Lesch (1984).
23. Langley (1901a, p. 224). On the development of the neurone theory see Robinson (2001, pp. 1–30).
24. Langley (1901a, pp. 224–5, 229).
25. Oliver and Schäfer (1895, pp. 239, 245–7).
26. Lewandowsky (1899) and Boruttau (1899, pp. 109–11).
27. Langley (1901b). In 1946 Ulf Svante von Euler provided persuasive evidence that noradrenaline was the transmitter substance of sympathetic nerves. For the history of chemical neurotransmission, see Valenstein (2005), Robinson (2001), Bennett (2000) and Dupont (1999); and for the relationship between transmitter and receptor research, Chapter 4 below.
28. Langley (1903a, pp. 6–7). On Ehrlich's toxin-antitoxin studies see Chapter 1 above.
29. Elliott (1905, p. 402). On the isolation of adrenalin see Tansey (1995a).
30. Elliott (1904, p. xxi).
31. Brodie and Dixon (1904, pp. 500–1) and Elliott (1905, pp. 431–2).

32. Brodie and Dixon (1904, pp. 497–8).
33. Elliott (1905, pp. 434–8). On the identification of acetylcholine as the transmitter substance in these other synapses, see Robinson (2001), Dupont (1999) and Tansey (1995b).
34. Elliott (1905, p. 467) acknowledged his 'indebtedness for advice' to Langley, but also to Gaskell and H. K. Anderson.
35. Langley (1905b, pp. 374–80).
36. Langley (1905b, pp. 380–93).
37. Ibid., pp. 393–9.
38. Ibid., p. 399 (emphasis in the original).
39. Ibid., pp. 399–400.
40. Ibid., p. 411.
41. Langley (1906a, p. 181).
42. Parascandola and Jasensky (1974, pp. 210–11) and Chapter 1 above.
43. Langley (1905b, p. 412).
44. Ibid., pp. 405–10, 413 and Langley (1906a, p. 193).
45. Elliott (1907, p. 442).
46. Dale (1961).
47. Langley (1906a).
48. Langley (1906b).
49. Langley (1907b) and, for a full account, Langley (1907–09).
50. In 1908 Magnus became professor in Utrecht, where he founded the first pharmacological institute of the Netherlands. See Magnus (2002). For his collaboration with Langley in 1905 in the Cambridge Physiological Laboratory, see Magnus (2002, pp. 133–4) and Langley and Magnus (1905).
51. Magnus (1908).
52. Ibid., p. 106 and Magnus (2002, pp. 66, 133, 138–41).
53. Magnus (1908, p. 107).
54. Magnus (1908, pp. 108–12).
55. Langley (1907–09, vol. 37, p. 299).
56. Fühner (1907) and Langley (1907–09, vol. 37, pp. 298–9).
57. Dixon and Hamill (1909). See also Chapter 3 below.
58. Barger and Dale (1910, p. 56). Dale had been a student of Langley. For further details see Chapter 4 below.
59. Boehm (1895).
60. Straub (1907).
61. Straub (1908, pp. 11–13). For a full discussion of Straub's criticism of the receptor theory see Chapter 3. On the early history of studies into the relationship between chemical structure and physiological effects of drugs see Bynum (1970) and Parascandola (1974).
62. Langley (1907–09, vol. 39, pp. 291–2).
63. Langley (1907–09, vol. 39, pp. 294–5). On Ehrlich's chemotherapeutic experiments on trypanosomes see Parascandola and Jasensky (1974, pp. 216–20), Silverstein (2002, pp. 130–1) and Weatherall (1990, pp. 58–60).
64. Langley (1907–09, vol. 39, p. 295).
65. Hill (1909). See also Chapter 5 below.
66. Langley (1910; 1914, p. 106; 1918).
67. Langley (1921, p. 44).
68. Compare Geison (1978, pp. 331–55).

69. Compare Sherrington (1953). Thanks to Dr John Forrester, Edinburgh, for this reference.

3 Receptors and Scientific Pharmacology I: Critics of the Receptor Idea and Alternative Theories of Drug Action, *c.* 1905–35

1. Bickel (2000, pp. 13–27), Maehle (1999; 2002) and Warner (1997, pp. 258–83). Concerning the history of pharmacology in general, see furthermore Stille (1994) and Müller-Jahncke and Friedrich (1996).
2. Nurmand (2004, esp. pp. 152–4).
3. Binz (1890, esp. pp. 62–74). Besides Buchheim and Schmiedeberg, Binz was one of the most important pharmacologists in the second half of the nineteenth century. See Bickel (2000, p. 69). For Buchheim, see Bruppacher-Cellier (1971).
4. Buchheim (1876, esp. pp. 265–9).
5. Ibid., p. 272.
6. For Buchheim's approach see Schmiedeberg (1912, esp. pp. 18–54).
7. Meyer (1922, esp. pp. 2, 4–5).
8. Brunton (1893, p. 35) and Bynum (1970).
9. Meyer (1922, pp. 4, 9).
10. Parascandola (1992).
11. Straub (1897).
12. Back (1986, pp. 23–7, 52). Walther Straub to the Principal of the University of Freiburg, Freiburg, 1 January 1908; Dean of the Medical Faculty to the Principal of the University of Freiburg, 3 January 1908; Ministry of Education (*Minister der Justiz, des Kultus und des Unterrichts*) to the Senate of the University of Freiburg, 25 January 1908 (two letters), in B 24/3743 (Personalakte), Straub, Walther, von Augsburg (n.p.), UAF. See furthermore the negotiations between the Medical Faculty of the University of Freiburg, the Philosophical Faculty and the Ministry of Education in the German Country of Baden, in B1/1225 (Lehrstuhl) Generalia. Medizinische Fakultät. Diener und Dienste. Den Lehrstuhl für Pharmakologie betr. (1907–27), UAF; Starke (2004b) and Eyer (2004).
13. Gremels (1947, esp. pp. 4–5). Gremels's paper includes a bibliography of Straub's publications (pp. 5–12). See also Eulner (1970, p. 136) and Heubner (1944, p. 139). Concerning the international reputation of Straub, see Gremels (1947, p. 3), Forst (1974, esp. pp. 1171, 1173) and Haffner (1944, esp. p. 343).
14. See Starke (1998, esp. pp. 17, 18, 20).
15. 'Straub, Walther' (1998) and Forst (1974, p. 1174); British Pharmacological Society. General Minutes, Box 1, Vol. I, 1931-5, SA/BPS/A 1, CMAC. See also Bynum (1981a).
16. Forst (1974, p. 1173). This is an ironic allusion to the medical term 'strabismus convergens' (that is, horizontal squint).
17. Back (1986, pp. 160–1), Nana Djiepmo (2005, pp. 6–7) and Döring (1996). Concerning 'experimental systems' see Rheinberger and Hagner (1993).

18. Back (1986, pp. 53–4, 165–6), Rost (1947, esp. pp. 202–3) and Stroomann (1960, pp. 36–7).
19. Straub (1920, pp. 246–7, 270–1, 300–1, here p. 6): 'Die heutige Pharmakologie ist eine experimentelle Wissenschaft, die von den Aenderungen in Zuständen und Vorgängen im lebenden Organismus handelt, wie sie durch chemische und physikalische Eingriffe hervorgerufen werden ... So ist sie angewandte Physiologie und arbeitet mit dem Mittel dieser, dem Experiment in physikalischer und chemischer Richtung. Sie ist experimentelle Medizin und ein Teil von dem, was man Pathologische Physiologie nennt.'
20. See, for example, Straub's collaboration with the clinician Krehl: Straub and Krehl (1919).
21. The Zoological Research Institute in Naples was founded by the German biologist Anton Dohrn and opened in 1874. The foundation originated from the attractiveness of sea life as models for the general biology of animal and man. Dohrn found Naples attractive because of the multitude of organisms in the sea off the coast and also because of its site and international reputation. In the next hundred years, the institute became one of the most important centres for biological research. For its history see Müller (1976), Partsch (1980) and Simon (1980); here especially Flad-Schnorrenberg (1980).
22. Back (1986, p. 53). For Straub's research work at the Zoological Research Institute in Naples, see Straub (1903, 1905, 1907). The second paper was an expanded version of his short preliminary account of 1905.
23. Straub (1907, pp. 133–4, 135; 1905, p. 303).
24. Straub (1905, p. 303).
25. Straub (1907, p. 138).
26. '... dass die fraglichen Vorgaenge sich doch wahrscheinlich in der physikalischen Grenzschicht jedes Zellindividuums abspielen'; see ibid., p. 139.
27. Boehm (1895, esp. pp. 16, 17–18).
28. Straub (1907, pp. 139–42).
29. Ibid., pp. 143–4.
30. Ibid., pp. 144–5, 149.
31. Maehle (2004b, pp. 171–2) and Chapter 2 above.
32. Clark (1933, pp. 189–90). Clark also described the reaction of Straub, especially to the opponents of his theory (pp. 190–5). See also Forst (1974, p. 1171) and Back (1986, pp. 158–9).
33. Straub (1908, esp. pp. 5–6, 11–16); also compare with: Straub to the Principal of the University of Freiburg, 4 February 1908, in B 24/3743 (Personalakte), Straub, Walther, von Augsburg (no pagination), Universitätsarchiv der Universität Freiburg.
34. Straub (1908, pp. 24–5; 1938, pp. 90–1, esp. p. 91). There were different views in the scientific community on the relationship of physiology and pharmacology, but this is another field, which cannot be discussed here.
35. Straub (1910, esp. p. 393).
36. Straub (1912).
37. Ibid., p. 3.
38. Ehrlich (1902, pp. 570–95, 596–618).

39. Straub (1912, p. 4): 'Ich halte diese Erklärung in ihrer generellen Fassung für zu weitgehend und unstatthaft und schließlich auch für unfruchtbar.'
40. Ibid., p. 4: 'Wollte man für diese, wie gesagt, zahlreichen Fälle mit "Chemorezeptoren" operieren, so verwandelt man nur dabei das eine Wunder in ein anderes. Die Existenz von Chemorezeptoren für Gifte soll nicht geleugnet werden, aber generell ist sie nicht, und damit kann auf ihnen auch keine weitfassende Theorie basieren.'
41. Ehrlich (1909, esp. pp. 173–4).
42. Straub (1912, p. 5).
43. Ibid., p. 5–6.
44. Maienschein (1990, pp. 370–1).
45. Straub (1912, pp. 23–5). For Straub and his theory see also Starke (2004a, pp. 3–15).
46. Alsberg (1921), Overton (1901) and Meyer (1899).
47. Clark (1933, p. 192). See as examples for the support of Straub's potential poison theory Wertheimer and Paffrath (1925, esp. pp. 254, 263, 266–8), Neukirch (1913, here esp. pp. 154, 169), Kuyer and Wijsenbeek (1913, esp. pp. 36–8), Cook (1926/27), Meyer (1908) and Kretschmer (1907). See also Parascandola (1986).
48. See as an example Rentz (1929) and Nanda (1931). It is not possible to give here a full account of this discussion.
49. Straub (1937).
50. Straub (1916, p. 144; emphasis in the original): 'Spekulationen über *unmittelbare* Beziehungen zwischen Konstitution und Wirkung und Natur der Verbindung von Gift und lebendem Organ sind wie immer müßig.'
51. Straub (1931b, here p. 45).
52. Ibid., p. 47.
53. Ibid., p. 45.
54. Straub (1931a).
55. Straub (1936).
56. Straub (1911; 1915; 1919) and Straub and Amman (1940).
57. Straub (1937). See also Gremels (1947, pp. 7–12).
58. Dale (1926a, pp. xix–xxvii) and Abel (1926). For Cushny see furthermore the literature quoted below as well as the unpublished biography by his daughter Helen Macgillivray, in: Cushny, Professor Arthur Robertson, FRS (1866–1926) PP/ARC, Wellcome Contemporary Medical Archive Center. This typescript manuscript contains mainly information about his family life.
59. See also Prüll et al. (2003, esp. pp. 24–6), Bynum (1981a, p. 5) and comment by Alfred Joseph Clark, in Cushny (1926a, esp. p. 457).
60. See respective remarks in Cushny (1901, esp. p. 7).
61. In Edinburgh, Cushny succeeded Sir Thomas Fraser, who held the joint chair of materia medica and clinical medicine, see Dale (1926a, p. xxv). We cannot agree with the opinion of Sir Henry Hallett Dale, who saw Cushny's interest in practical drug therapy as a contrast to the Schmiedeberg school (p. xx).
62. Ibid., pp. xx, xxiii; statement of Walter Dixon, in Cushny (1926b) and Cushny (1901, p. 9).
63. See Cushny (1901, p. 7; 1926b, quotation on p. 519) and Dale (1926a, p. xxvi). See also Cushny (1926a) and Abel (1926, p. 269).

64. Dale (1926a, pp. xx, xxi–xxv).
65. These are substances with basically the same chemical structure, which show a different polarization in light. Ibid., p. xxv.
66. Abel (1926, p. 277).
67. Dale (1926b, esp. p. 7).
68. Cushny to Abel, 22 February 1909, in Cushny, Prof. Arthur Robertson, PP/ARC/D1, CMAC.
69. Parascandola (1975). Quotation from Abel (1926, p. 280).
70. Parascandola (1975, pp. 150–3). For Cushny's experimental setting, see Cushny (1908, esp. pp. 136–8).
71. Parascandola (1975, pp. 156–7).
72. Ibid., pp. 156–7; Letter from Cushny to Abel, 29 October 1914, in Cushny, Prof. Arthur Robertson, FRS (1866–1926), PP/ARC/D1, CMAC.
73. Parascandola (1975, pp. 157–61).
74. Cushny (1901, pp. 20, 24). The same phrases can be found in the first edition of 1899.
75. Cushny (1903, p. 20; 1907, p. 20).
76. Cushny (1903, p. 20).
77. Cushny (1910a, p. 5).
78. Ibid., pp. 20–1, 25, 261, 712–13; see the quotation on p. 25.
79. Cushny (1915b, pp. 19–20, see the quotation on p. 20).
80. Cushny (1918, pp. 19–20; 1924, pp. 19–20).
81. Comment by Ernest H. Starling, in Cushny (1926a, p. 456).
82. This is supported by the next two editions of Cushny's textbook, which appeared after his death. The pharmacologists J.W. Edmunds and J.A. Gunn kept to Cushny's intermediate position but clearly strengthened the chemical point of view: Cushny (1928, esp. pp. v, 25; 1934).
83. Cushny to Abel, 3 November 1915, in Cushny, Prof. Arthur Robertson, FRS (1866–1926) PP/ARC/D1, CMAC.
84. Alsberg (1921, esp. pp. 327–80) and Overton (1901, esp. pp. 53–78).
85. Meyer (1899, esp. pp. 112–13; 1901, esp. p. 338). Remarkably, Meyer saw his research not as an opposing theory to Ehrlich's work but rather as an addition or improvement of certain aspects. See ibid., p. 343. See also with connections to colloidal chemistry, Overton (1899).
86. Typescript biography by Cushny's daughter, Helen Macgillivray, in Cushny, Prof. Arthur Robertson, FRS (1866–1926) PP/ARC/A4, CMAC.
87. Traube (1919, here p. 178). This was an application of his theory to pharmacology: Traube (1910; 1911).
88. Traube (1919, pp. 178–80; see also p. 185). With the concentration on surface energies, Traube relied to a great extent on the writings of Warburg, Battelli and Stern.
89. Ibid., pp. 181–7.
90. Ibid., pp. 187–91.
91. Ibid., pp. 191–3, 195. Traube supported this demand with a hint towards Fraenkel's textbook of drug synthesis, which was a forerunner in his view.
92. Mazumdar (1989, esp. p. 13).
93. Ibid., pp. 13–14, quotation on p. 14. See also especially for Landsteiner and Ehrlich, Mazumdar (1995).
94. Bayliss (1915, pp. 1–5, 11–12).

95. Letter Straub to Bayliss, 13 May 1909, 9 June 1909, in Bayliss, Sir William, FRS (1860–1924), GC/223/A2, CMAC; Letter Rudolf Boehm to Bayliss, 12 June 1911, ibid.
96. Gunn (1949, p. 231), Bynum (1981a, pp. 9–10), and Rolleston (1931); in Dixon, Walter Ernest, FRS, FRCP, OBE, MD (1870–1931) and Myers (George) Norman, FRCP, MD, PhD (1898–1981), GC/155/A1, CMAC.
97. Dixon and Hamill (1909, esp. p. 314).
98. Ibid., p. 335.
99. Ibid., pp. 331–6. For Dixon's approaches to pharmacological research, see also Dixon, Myers, GC/155/B4, CMAC. Although the materials of Dixon and Myers are partly intermingled in this file, Dixon's style of experimentation can be well reconstructed.
100. Dixon (1906).
101. Ibid., pp. 5–7, 16–17.
102. Dixon (1908).
103. Dixon (1913, p. v).
104. Ibid., pp. 6, 69–71.
105. Dixon (1915, p. v).
106. Ibid., pp. 70–3.
107. Ibid., p. 100. Dixon did not mention Elliott, but mainly the works of Keith Lucas in connection with the description of the 'myo-neural substance', p. 101.
108. Dixon (1921, p. 22).
109. Dixon (1925, p. v).
110. Ibid., p. 22.
111. Dixon (1929).
112. Compare Whitla (1903, p. 109; 1910, p. 103; 1915, pp. 103–4; 1923, p. 518).
113. Other examples for an intermediate position, pointing out the importance of both approaches, are Charles Wilson Greene, professor of physiology and pharmacology at the University of Missouri (Greene 1914, esp. pp. 6–7) and W. Storm van Leeuwen, pharmacotherapeutical institute at the University of Leiden (Storm van Leeuwen 1924, esp. p. 31), as well as J.A. Gunn, professor of pharmacology at the University of Oxford (Gunn 1931, 1932).
114. Sollmann (1901, Preface).
115. Ibid., p. 129.
116. Ibid., pp. 129–30, 398.
117. Sollmann (1922, pp. 73–4).
118. Ibid., pp. 74–5.
119. Sollmann and Hanzlik (1928, p. 169).
120. Ibid., pp. 115, 170.
121. For Fechner and Weber and the history of the Weber-Fechner Law, see Heidelberger (1996, p. 238), Arendt (2001, pp. 2–14; 1999) and Lennig (1994a, pp. 159–74; 1994b). See also Clark (1933, p. 197).
122. Ibid., p. 197.
123. Fechner also influenced, for example, the development of twentieth-century psychology and psychiatry on the basis of his contacts with the psychologist Wilhelm Wundt (1832–1920), who until the end of his life remained in touch with the influential psychiatrist Emil Kraepelin (1856–1926), see Hoff (1995, esp. p. 261) and Arendt (2001, p. 12).

124. Matthews (1931).
125. Ibid., pp. 65–6.
126. See ibid., pp. 64, 67, 108–9.
127. Eulner (1970, p. 659).
128. Pütter (1918).
129. Ibid., pp. 201, 203–5, 214–15, 230, 250, 224, 228, 251–61.
130. Wald (1991).
131. Hecht (1931).
132. Ibid., pp. 253–7, 270, 307–12, 316, 351, 362.
133. Clark (1933).
134. Ibid., p. 195.
135. Ibid., p. 195. See also Lucae (1998, pp. 145, 215–16).
136. Schulz (1888, esp. pp. 540–1).
137. Lucae (1998, p. 216) and Jütte (1996, esp. pp. 34–8).
138. For holism and constitutionalism, especially in the third decade of the twentieth century, see Lawrence and Weisz (1998), particularly the introduction of both editors on pp. 1–22, and Prüll (1998), Timmermann (2001), Harrington (1996) and Klasen (1984).
139. Krügel (1984).
140. Martius (1923, pp. 1005–6).
141. Ibid., p. 1006; Lucae (1998, pp. 146, 201, 216) and Clark (1933, p. 195).
142. Ibid., p.195.
143. Handovsky et al. (1923, esp. pp. 273, 281–4).
144. Dannenberg (1930).
145. Clark (1933, see the quotation on pp. 196–7).
146. Concerning chemistry and medicine in the nineteenth century, see Bynum (1994, pp. 118–23).

4 Receptors and Scientific Pharmacology II: Critics of the Receptor Idea and Alternative Research Strands: the Transmitter Theory, c. 1905–35

1. See Robinson (2001, pp. 1–47), Rapport (2005) and Clarke and Jacyna (1987).
2. Robinson (2001, pp. 49–56). For the history of neurotransmission see also Bacq (1983), Dupont (1999) and Valenstein (2005).
3. For the life of Dale, see Feldberg (1970), Bynum (1981b, pp. 104–5) and Tansey (1990). For his own comments on some of his important publications, see Dale (1953). Dale's importance in the interwar period becomes visible when looking at his activities in the Medical Research Council and especially at his involvement in the introduction of new drugs in Britain: Elliott, Professor Thomas Renton, CBE, DSO, FRCP, FRS (1877–1961), GC/42, CMAC, here esp. GC/42/13 and 15. For the Wellcome Physiological Research Laboratories see also Tansey (1989).
4. See for Barger: Clark (1940, pp. 260–1) and Dale (1941).
5. Tansey (1990, pp. 360–90) and Bynum (1981b, pp. 105–6).
6. Bynum (1981b, p. 106), Tansey (1990, pp. 119–74, 238–50, 253–61) and Dale (1961). The most important of Elliott's papers in our context is Elliott (1905).

7. Tansey (1990, pp. 277–80) and Bynum (1981b, p. 106).
8. See Valenstein (2005, pp. 29–50) for a description of Dale's career development.
9. For Loewi see Lembeck and Giere (1968) and Geison (1981). See also Bericht der Berufungskommission zur Wiederbesetzung des Lehrstuhls für Pharmakologie an der Universität Graz, 8.2.1909, in: Univ. Prof. Dr. Otto Loewi, geb. 3.6.1873. Personalakte, Archive of the University of Graz, pp. 7–8, here p. 7. See furthermore F. Brücke and Otto Loewi, in Jahrbuch der Heilmittelwerke Wien, 1961 (engl. Translation), Material Collection Otto Loewi, Archive of the University of Graz, pp. 731–53, esp. pp. 734–5; Otto Loewi (1960, p. 6) and Lembeck et al. (2004).
10. See the following important papers: Loewi (1921; 1922; 1924a; 1924b; 1924c), and Loewi and Navratil (1926a). See also Valenstein (2005, pp. 51–7).
11. Tansey (1990, pp. 318–59). It is not possible to give a full account of the works of Dale and Loewi here, and the reader is advised to use the quoted literature for more detailed information. The intention was to give a short description of the careers of both scientists as a background to their impact on the development of the receptor concept. For Dale's work on acetylcholine see esp. Tansey (1991; 1995b). For the history of Dale's and Loewi's Nobel Prize, see Valenstein (2005, pp. 68–88).
12. Dale (1965, p. 412). Dale at least acknowledged Langley's influence on his basic direction of research. For Dale's missing acknowledgement of Langley's influence on his own work on chemical interaction, see Tansey (1990, p. 275). See also the following early papers showing Langley's influence: Dale (1899/1900; 1901/02).
13. Valenstein (2005, p. 46) and remark by Edgar Lord Adrian on Dale in 1955, quoted in Valenstein (2005, p. 46).
14. See again the speech of Lord Adrian on the occasion of Dale's 80th birthday in 1955 (Tansey 1990, p. 309); Henry Hallett Dale, Chemical Structure and Physiological Action. Herter Lectures 1919, III, manuscript, p. 17, in Dale, Professor Henry Hallett, OM, GBE, FRS (1875–1968), PP/HHD/15, CMAC. Examples for Dale's style of experimentation: Dale (1905–1906), Dale and Gasser (1926). See furthermore Dale's decisive experiments on the cholinergic component of ergot in 1913: Dale, Professor Henry Hallett, OM, GBE, FRS (1875–1968), PP/HHD/2, CMAC.
15. Tansey (1990, p. 272).
16. Barger and Dale (1910, p. 56). See also Chapter 2 above.
17. Barger and Dale (1910, p. 58).
18. Letter Dale to Elliott, 11 December 1913, in Elliott, Professor Thomas Renton, CBE, DSO, FRCP, FRS (1877-1961), GC/42, CMAC. See also Tansey (1990, p. 275).
19. Dale (1906, pp. 173, 203).
20. Tansey (1990, pp. 256–7).
21. Barger and Dale (1910, pp. 54–5).
22. Dale (1935b, p. 330) and Tansey (1990, p. 267).
23. Barger and Dale (1910, p. 56).
24. Dale (1920, pp. 376–7, 379–80).
25. Dale (1921, pp. 9, 10–14).
26. Dale (1923, pp. 380, 390).

27. Dale (1932; 1934, see esp. the quotation on p. 347; 1935a).
28. 'Beta-receptors' are a certain receptor type that was investigated only in 1948 by Raymond Ahlquist. Chapter 6 will discuss this topic.
29. Tansey overestimates Dale's contribution to the development of the receptor idea with this statement, which is based on a too general treatment of Dale's relationship to the research on receptors. The reason is that Tansey does not sufficiently distinguish between the strands of transmitter and of receptor research, and this prevents a deeper analysis of Dale's attitudes. See Tansey (1990, p. 315). See also Feldberg (1970, p. 128). The idea that Dale neglected the receptor concept in his work was mentioned by Sir James Black in 1976, when he wrote that 'Dale's attitude seems to have had a powerful effect in delaying the introduction of the idea of receptors into pharmacological teaching and his impact was still dominant when Ahlquist's paper appeared in 1948.' Black mainly referred to Dale's failure to support the receptor concept, whereas Dale also actively tried to push back initiatives by the supporters of the receptor concept. Furthermore, Black identified another threat to the receptors in the work of the American physiologist Walter B. Cannon and his student Arturo Rosenblueth, who developed a new theory of the function of the sympathetic nervous system in 1932 (see Chapter 6). See Black (1976, p. 12).
30. Loewi (1902a, pp. 325–6).
31. Loewi (1902b, p. 437).
32. Loewi and Solti (1922).
33. Loewi (1922, p. 212).
34. Loewi and Navratil (1926a, pp. 679–80). See furthermore Loewi (1927b, p. 2176) and Engelhardt and Loewi (1930, p. 4).
35. Loewi and Navratil (1924, p. 123).
36. Loewi and Singer (1924).
37. Compare, especially, as a summarizing comment Loewi (1927a, pp. 93–4). See also Häusler and Loewi (1925, p. 275), Loewi (1926, pp. 1075–6), Dietrich et al. (1926, 1927), Häusler and Loewi (1926) and Dietrich and Loewi (1927).
38. Compare Loewi and Navratil (1926b) and Loewi (1931; 1965).
39. Loewi (1929). See the summary in Loewi (1930).
40. Compare, for example, Fröhlich and Loewi (1908); Langley is quoted in Loewi (1908, p. 69).
41. Loewi (1910, p. 275).
42. Loewi (1937a, p. 8).
43. '... Wir kennen den letzten Wirkungsmechanismus der Ueberträger so wenig wie den andrer Alkaloide ... Der Angriffspunkt der Stoffe ist nach alldem irgend ein chemisch oder chemisch-physikalisch charakterisierter Bestandteil des Erfolgsorgans, nämlich die Zelle selbst. An dieser wirken die Stoffe, nach allem, was wir wissen, auf die Zelloberfläche und zwar nur von außen. ... Daß der Wirkungserfolg ein und desselben Stoffes hauptsächlich an der Stelle eintritt, wo er freigemacht wird, daß also gerade dort die Zellen für seine Wirkung empfindlich sind, ist eine Teilerscheinung jener nicht kausal, sondern nur teleologisch begreiflichen, spezifischen Empfindlichkeit gegenüber bestimmten chemischen Stoffen, die wir allenthalben im lebenden Organismus als eine der Unterlagen seiner Funktions- und damit Existenzfähigkeit antreffen ...' See Loewi (1937b, p. 854).

44. Fred Lembeck and Walter Kukovetz's remark that Loewi 'unequivocally proved' the effect of the *'Vagusstoff'* and the *'Acceleransstoff'* on the 'receptive substances' of the heart muscle is a somewhat unfortunate phrase as it suggests Loewi's full acceptance of the receptor concept as a part of research on neural transmission. The same is true for Lembeck's remark that Loewi could show that atropine would 'block' the receptors for acetylcholine. See Lembeck and Kukovetz (1985, p. 267). On pharmacologists' difficulties with the receptor concept, see Valenstein (2005, pp. 170–1).
45. Ewald (1944, pp. 126–7) and Erbslöh (1958, pp. 724–7).
46. Hoch et al. (2001) and Vincent et al. (2001).
47. Dale never saw himself as a disciple of Langley, although he was educated in Langley's laboratories.

5   Quantitative Arguments for the Existence of Drug Receptors and the Development of the Receptor Occupancy Theory, *c.* 1910–60

1. Clark (1933, p. 1).
2. For brief accounts see also Robinson (2001, pp. 143–52) and Colquhoun (2006); and on Clark see also Parascandola (1982; 1986, pp. 148–53). An earlier version of this chapter has been published by Maehle (2005).
3. See also Chapter 2 above.
4. Hill (1909).
5. Straub (1907). See also Straub (1912).
6. Hill (1926, p. 279) and Katz (1978, p. 77).
7. Barcroft and Hill (1910), Hill (1910) and Katz (1978, p. 78).
8. Hill (1965).
9. Hill (1926, pp. 276–8). The 'lock and key' metaphor had been introduced in 1894 by the Berlin chemist Emil Fischer (1852–1919) in describing the spatial fit between enzymes and their substrates. On the influence and scientific context of this metaphor see Cramer (1997) and Ramberg (2003). See also Chapter 1 above.
10. On Dixon see Clark (1932) and Chapter 3 above. For Clark's biography see Hazell Clark (1985) and Verney and Barcroft (1941).
11. Dixon and Hamill (1909).
12. Cushny (1910b).
13. See Parascandola (1975), and Chapter 3 above.
14. Compare Straub (1910) and Clark (1912; 1913a). Clark had spent some months at the University of Marburg in Germany during 1907 (see Hazell Clark 1985, p. 5). Straub had been teaching at Marburg in 1905–06, but moved on to Würzburg in 1906–07.
15. Clark (1912; 1913a; 1913b). Clark's extensive work with the isolated frog heart preparation led him also deeply into the field of comparative cardiac physiology (see Clark 1927a). In Edinburgh he collaborated in this field among others with the biochemists P. and M. Grace Eggleton of the department of physiology. See Clark et al. (1938).
16. See Hazell Clark (1985, pp. 5–6, 14).

17. Loewi (1921; 1922), Loewi and Navratil (1926a) and Tansey (1995b). For Loewi's research on transmitter substances and his view on receptors, see Chapter 4 above.
18. Clark (1926a).
19. Ibid., pp. 534–46.
20. Ibid., p. 545.
21. Straub (1907, p. 148) and Langley (1909/10, p. 292). See also Chapter 2 above.
22. Cushny (1915a, pp. 448–9).
23. For Gaddum's biography, see Feldberg (1967).
24. Gaddum (1926, p. 142).
25. Ibid., pp. 149–50.
26. Clark (1926b).
27. Ibid., p. 555.
28. See Clark (1912; 1913b; 1922) and De Burgh Daly and Clark (1921).
29. Clark (1926c).
30. Loewi and Navratil (1926a, p. 688). See also Chapter 4 above.
31. Clark (1927b, p. 314).
32. Clark (1933, p. 133).
33. Clark (1937a, pp. 25–6).
34. Feldberg (1967, p. 61).
35. Gaddum (1937).
36. Clark (1937a, pp. 216–17). See also Clark and Raventós (1937, p. 391).
37. Clark (1937b, p. 584) and Straub (1937).
38. Compare Verney and Barcroft (1941, p. 984). See also Dunlop (1940–42), Gaddum (1941) and Burn (1942).
39. Clark (1929, p. vi).
40. Ibid., pp. vi–vii (reprinted Preface to the first edition). There were also several posthumous editions of Clark's *Applied Pharmacology*, carrying his name until the 9th (reprinted) edition of 1961; see Wilson and Schild (1961).
41. See Clark (1938).
42. Hazell Clark (1985, pp. 44–5).
43. Since 1938 Gaddum had been professor of pharmacology in the College of the Pharmaceutical Society, London. See 'Chair of Materia Medica', *University of Edinburgh Journal*, 11 (1940–42), p. 237.
44. On Ing see Schild and Rose (1976).
45. Dale (1943, p. 320).
46. Dale (1943, p. 321) and Ing (1943).
47. Ing (1943, pp. 373, 379) and Schild and Rose (1976, p. 247).
48. Gaddum (1943).
49. On the history and impact of this research strand see Lesch (2007).
50. The history of this finding, which was mostly due to the work of Donald Devereux Woods (1912–64) and Paul Fildes (1882–1971) at the Middle-sex Hospital, London, has been briefly summarized by Weatherall (1990, pp. 152–4).
51. Ariëns (1954).
52. Stephenson (1956, p. 380).
53. Clark (1933, p. 61).
54. Clark (1937a, p. 193).

55. Stephenson (1956, p. 391). See also the similar explanation given by Nickerson (1956).
56. Stephenson (1956, p. 392).
57. Katz and Thesleff (1957).
58. Paton (1961). For Paton's biography see Rang and Walton (1996).
59. Rang and Walton (1996, p. 304).
60. Ariëns (1964).
61. De Jongh (1964, p. xvi).
62. As Gaddum wrote in February 1960 in his draft preface to Ariëns's handbook: 'Pharmacologists that commonly spend much of their time instructing medical students and general pharmacologists are rare. They must be widely read and have good memories. They must have the energy to collect data from many sources and arrange them in an orderly way, so that general principles appear through the fog of irrelevant facts. A.J. Clark could do this, and E.J. Ariëns has shown that he can do it too.' Gaddum, Sir John, FRS (1900–65), GC/213/A.1, Ariëns, Prof. E.J., January 1960–March 1963 (Box 1), CMAC. See also Gaddum (1964).

6 The Dual Adrenalin Receptor Theory of Raymond P. Ahlquist (1914–83) and its Application in Drug Development between 1950 and 1970

1. Cooter and Pickstone (2000, pp. xiii–xix).
2. Carrier and Shlafer (1984, p. 41), Little (1987, p. 583); James H.R. Sutherland, History. Department of Pharmacology and Toxicology. School of Medicine. Medical College of Georgia, 1988, typewritten manuscript, p. 10, courtesy of Lois T. Ellison, Medical Historian in Residence, Augusta; J. H. R. Sutherland, G. O. Carrier, L. M. Greenbaum, Obituary Raymond P. Ahlquist, 26 July 1914–15 April 1983, manuscript, in Dean's Office Newsletter, MCG (Medical College of Georgia), 8 (1983), No.15, 29 July, one page, in Papers of Lowell Greenbaum, Augusta.
3. Carrier (1986, p. 206).
4. Ibid., pp. 208–9; Barger and Dale (1910).
5. Ahlquist et al. (1947) and Ahlquist and Woodbury (1947).
6. Adrenalin itself could not be used because of its cardiovascular side-effects, see Carrier (1986, p. 209).
7. Ibid., pp. 205–6; Carrier and Shlafer (1984, pp. 41–2).
8. 'Isolated' means removed from the animal body, 'intact' means left within the animal body.
9. Ahlquist (1948, pp. 587–95). The total of 23 different tissues results from the fact that Ahlquist did not use every animal species for every organ system. See also Robinson (2001, p. 153). Table I of Ahlquist's paper, which summarizes the results, is reduced to a simplified version without any comment; Carrier (1986, p. 211). Carrier reduces the number of investigated substances to five, leaving out levo-adrenaline, presumably for didactic purposes.
10. Ahlquist (1948, pp. 588, 591–5) and Carrier and Shlafer (1984, p. 43).
11. Ahlquist (1948, pp. 590, 596).

12. Ibid., p. 599. For Ahlquist's dual adrenergic receptor concept see also Carrier (1986, pp. 211–17).
13. Black (1976, p. 12), Dale (1954), Cannon and Rosenblueth (1933; 1937) and Wolfe et al. (2000).
14. Valenstein (2005, pp. 89–120, esp. pp. 111–14).
15. Ahlquist (1948, pp. 586–7, 596–9). In the course of investigations on Ahlquist's receptor theory during my research stay in Augusta in summer 2003, two laboratory books of Ahlquist turned up. The first one was entitled 'Sympathomimetic Amines. Book I. Compound and Reference Numbers. R. P. Ahlquist 1945'. It contains only formulas, reference numbers with company names and literature hints, but no remarks or descriptions of his experiments. The same is true for the second book, entitled 'Sympathomimetic Amines. Book II', which contains only an evaluation of the literature about the topic from the years 1945–6. Both books were saved by Richard E. White in 1983, when Ahlquist's room was emptied after his death. His papers were destroyed [Cay-Rüdiger Prüll].
16. Ahlquist (1948, p. 586) and Carrier (1986, p. 217). See also Valenstein (2005, pp. 115–16).
17. Speech of Raymond P. Ahlquist [draft] on receiving the Memorial Award in Therapeutics in 1973, typewritten manuscript, 6 pp., p. 3, in: Ahlquist, Raymond P., Folder, Department of Pharmacology and Toxicology, Medical College of Georgia, Augusta.
18. Concerning the relationship between Ahlquist and Hamilton, see ibid., p. 3; Carrier (1986, p. 217) and Carrier and Shlafer (1984, p. 42); Interview with Lois T. Ellison, MD, Medical Historian in Residence, Provost Emeritus, Professor Emeritus of Medicine and Surgery, formerly Director of the Cardiopulmonary Lab of the Medical College, Georgia, Augusta, on 9 September 2003, Transcript part 1, pp. 1–2, 6–7.
19. Carrier (1986, pp. 218–20) and Shanks (1984, pp. 39–40).
20. Levy and Ahlquist (1957, see the quotation on p. 420); speech of Raymond P. Ahlquist [draft], on receiving the Memorial Award in Therapeutics, 1973, p. 4.
21. Ibid. In 1962, Levy became professor of pharmacology at the University of Texas, Galveston, see James H. R. Sutherland, History. Department of Pharmacology and Toxicology. School of Medicine. Medical College of Georgia, 1988, typewritten manuscript, 10 pp., p. 2, courtesy of Lois T. Ellison, Medical Historian in Residence, Augusta.
22. See the short account of Ahlquist's work on the adrenergic system and especially on β-blockers after 1948 in Ahlquist (1980, pp. 103–4). In 1961, Ahlquist and Levy even developed a multi-organ testing system to study adrenoreceptor blocking substances, see ibid., p. 103.
23. Raymond P. Ahlquist, draft of a speech, no date, typewritten manuscript, pp. 11, p. 7, in: Ahlquist, Raymond P., Folder, Department of Pharmacology and Toxicology, Medical College of Georgia, Augusta.
24. Ahlquist et al. (1947, p. 287), Ahlquist and Woodbury (1947, p. 305) and Shanks (1984, pp. 40–1).
25. Powell and Slater (1958, p. 480, quotation on p. 486).
26. Ibid., pp. 480–8, quotations on pp. 486–7; Shanks (1984, p. 41).

27. This information stems from Moran himself, given in a personal communication to Shanks in 1982. Although given in hindsight, his description fits perfectly well with other facts and information obtained on the story of Ahlquist's theory. See Shanks (1984, pp. 41–2), Powell and Slater (1958, p. 480); personal communication of Armand M. Karow, PhD, who was assistant professor from 1968 and associate professor from 1971 at the pharmacology department, and finally president of Xytex Corporation, on 9 September 2003, Augusta.
28. Moran and Perkins (1958, p. 224).
29. Ibid., pp. 234–6 (discussion of the results). All quotations on pp. 234–5 (emphasis in the original).
30. Ibid., pp. 223–37.
31. Slater and Powell (1959, pp. 462–3).
32. Moran and Perkins (1961, pp. 192–4, 199–201).
33. Ibid., p. 200.
34. Shanks (1984, p. 39).
35. Moran and Perkins (1958, pp. 234–5; 1961, p. 200).
36. Slater and Powell (1959, pp. 462–3).
37. Personal communication of Shanks with Slater in 1982, quoted from Shanks (1984, p. 41).
38. Ibid., pp. 44–7. Carrier's statement that the dual adrenoceptor concept was 'unanimously accepted' in the 1960s is therefore misleading. See Carrier (1986, p. 219); speech of Raymond P. Ahlquist [draft], Memorial Award in Therapeutics, 1973, p. 4.
39. See Black (2004, p. 141).
40. Interview with Lois T. Ellison, 9 September 2003, Transcript part 1, p. 3.
41. Ibid., Transcript part 2, p. 3. Interview with Richard E. White PhD, former student of Ahlquist, now associate professor of the department of pharmacology and toxicology at Medical College of Georgia, Augusta, on 12 September 2003, p. 8. See also Little (1987, p. 584).
42. Sutherland, Carrier, Greenbaum, Obituary Raymond P. Ahlquist (see note 2 above).
43. Deriso (1988). Interview with Jerry Buccafusco, since 1979 assistant of Lowell Greenbaum (Ahlquist's successor in the chair of pharmacology), now director of the Alzheimer's Research Center, professor of pharmacology and toxicology, professor of psychiatry and health behavior, Medical College of Georgia, Augusta, 11 September 2003.
44. When Ahlquist met the assistant (Buccafusco) many years later, he said 'I must admit that I made a mistake. You are right with Clonidin. I take it myself and it works.': Interview with Jerry Buccafusco, September 8, 2003.
45. MCG Professor to receive Hunter memorial citation (1973); Dr. Raymond Ahlquist, noted researcher, dies (1983), p. 1; Symposium given for Dr. Ahlquist (1980); James H. R. Sutherland, History. Department of Pharmacology and Toxicology. School of Medicine. Medical College of Georgia, 1988, typewritten manuscript, 10 pp., p. 5, courtesy of Lois T. Ellison, Medical Historian in Residence, Augusta.
46. Information from Professor Lowell Greenbaum, successor of Ahlquist as head of the department of pharmacology and toxicology at Medical School of

Georgia, Augusta, in 1979, given on 26 January 2003; personal comments of Lowell Greenbaum, 7 September 2003.

47. Quotation taken from Taylor (1976).

48. Interview with Armand M. Karow, PhD, 9 September 2003, Augusta, p. 6; Information from Lowell Greenbaum, 26 January 2003; curriculum vitae of Raymond P. Ahlquist, typewritten manuscript, p. 7, in: material Lowell Greenbaum, Augusta; personal communication of Lowell Greenbaum, 7 and 9 September 2003, Augusta; Interview with Jerry Buccafusco, Augusta, 11 September 2003; Recognition justly due (1977). Acting chairman of the department of pharmacology between 1977 and 1979 was J. Malcolm Kling, see also Sutherland, History. Department of Pharmacology and Toxicology, pp. 5/6.

49. Personal communication of Gloria Greenbaum, Augusta, 9 September 2003; Interview with Lois T. Ellison, 9 September 2003, part I, p. 3.

50. Taylor (1976, first quotation here), Craig (1988), Guillebeau (1977, p. 4, second quotation here); Raymond P. Ahlquist, Draft of a speech, no date, typewritten manuscript, pp. 11, pp. 7, 8, in: Ahlquist, Raymond P., Folder, Department of Pharmacology and Toxicology, Medical College of Georgia, Augusta.

51. Profile (R. P. Ahlquist) (1975).

52. Ahlquist (1959), Ahlquist and Levy (1959), Levy and Ahlquist (1960, 1961). For Ahlquist's continuing interest in receptor research see Ahlquist (1976).

53. Ahlquist (1948, p. 598).

54. Ahlquist (1980, p. 105; 1962, p. 40).

55. Ahlquist (1973), see the first and third quotation here; Speech of Ahlquist [draft] on receiving the Memorial Award in Therapeutics in 1973, pp. 3, 5 (second quotation on p. 5); Ahlquist, Draft of a speech, no date, pp. 6, 9–11 (fourth quotation on p. 10); Lasker Laurels for Medical Researchers (1976), pp. 48–50, esp. p. 48; Interview with Jerry Buccafusco, 11 September 2003; Shanks (1984, p. 65) and Robinson (2001, pp. 154–5).

56. Interview with Richard E. White PhD, Augusta, 12 September 2003, p. 5; Speech of Ahlquist [draft] on occasion of the receipt of the Memorial Award in Therapeutics in 1973, p. 6. Even in 1967, Ahlquist pointed out that 'Although the alpha and beta receptors have achieved international usage it should be stressed that they have only interim value until the exact nature of the responsive mechanism for adrenergic agonists is discovered.' See Ahlquist (1967, p. 552). See also Ahlquist (1968).

57. Interview with Richard E. White PhD, Augusta, 12 September 2003, pp. 1–3; Interview with Jerry Buccafusco, 11 September 2003; Interview with Armand M. Karow, PhD, 9 September 2003, Augusta, pp. 4, 6. It seems also that the court trial brought against Ahlquist for his failure to grant tenure caused friction in the department and hampered laboratory work. See the Interview with Richard E. White PhD, Augusta, 12 September 2003, pp. 12–13; Information from Professor Lowell Greenbaum, Augusta, 26 January 2003.

58. Ibid., pp. 1, 4; Interview with Richard E. White PhD, Augusta, 12 September 2003, pp. 5, 10; Interview with Jerry Buccafusco, 11 September 2003; Nanette K. Wenger to Dorothea Ahlquist, 22 April 1983; W. T. Langston to Raymond P. Ahlquist, 4 November 1976; Paul Tumlin to Raymond P. Ahlquist, 22 October

1976, in: Ahlquist, Raymond P., Folder, Department of Pharmacology and Toxicology, Medical College of Georgia, Augusta.

59. Ahlquist (1980, pp. 104, 105, first quotation on p. 105); Interview with Armand M. Karow, PhD, 9 September 2003, Augusta, p. 3; Kass Wenger and Greenbaum (1984, pp. 419–21, second quotation on p. 420); Speech of Ahlquist [draft] on receiving of the Memorial Award in Therapeutics in 1973, p. 6 (third quotation); Interview with Richard E. White PhD, Augusta, 12 September 2003, pp. 2, 7, 10; Personal communication of Lowell Greenbaum, 7 September 2003, Augusta.

60. Interview with Richard E. White PhD, Augusta, 12 September 2003, p. 7 (quotation); Interview with Lois T. Ellison, 9 September 2003, Transcript part 2, p. 3; Interview with Jerry Buccafusco, 11 September 2003. Concerning teaching, see also the comments of his two collaborators Gerald O. Carrier and Marshal Shlafer, in Carrier and Shlafer (1984, pp. 42–4).

61. Interview with Armand M. Karow, PhD, 9 September 2003, Augusta, pp. 4–5; Kass Wenger and Greenbaum (1984, p. 419) and Witham (1977, pp. 637–8).

62. Quirke (2005).

63. Adam (2005).

64. Black and Stephenson (1962).

65. Shanks (1984, pp. 48–54).

66. Ibid., pp. 55–8; Black et al.(1964).

67. Shanks (1984, p. 58). For Black's research and research attitudes, see also Black (1988/89) and Quirke (2006).

68. Shanks (1984, pp. 63–5).

69. Ibid., pp. 65–6; Prichard (1964) and Prichard and Gillam (1964).

70. Shanks (1984, pp. 65–6), Prichard (1979, 1978).

71. Raymond Ahlquist to Lowell Greenbaum, 26 May 1980, in: Ahlquist, Raimond, P., Folder, Department of Pharmacology and Toxicology, Medical College of Georgia, Augusta; Stapleton (1997); Personal comment of Sir James Black, London, 11 June 2004; Black (2004, p. 141); Ahlquist (1958).

72. Sir James Black to Mrs Dorothea Ahlquist, Beckenham, 28 April 1983, in material Lowell Greenbaum. For the ß-blocker story see also Black (1993, pp. 418–25).

73. Deriso (1988).

74. Brian Prichard, Obituary Note on Raymond P. Ahlquist, in material Lowell Greenbaum; Stern Magazine, 1976, H.4, 15 January 1976, in: Ahlquist, Raymond P., Folder, Department of Pharmacology and Toxicology, Medical College of Georgia, Augusta; Craig (1988).

75. Another example for the importance of the practical application of an idea for its breakthrough in pharmacology are the ganglion blockers. See Timmermann (2006, pp. 157–74).

76. Robinson (2001, pp. 158–9, 161).

77. Breathnach (1978); British Pharmacological Society, General Minutes, Box 1, Vol. I, 1931–1935, SA/BPS/A3, A4, General Minutes, CAMC.

78. Personal comment of Sir James Black, London, 11 June 2004.

79. In his 1948 paper, Ahlquist gave Dale credit for the latter's contribution to receptor research in 1906, quoting Dale (1906) (Ahlquist 1948, p. 586). But Dale's paper only contains a minor footnote on 'receptors', giving the explanation that Elliott's term 'myoneural junction' is used instead. This

is basically no support for the receptor idea because the myoneural junction stands for the transmitter-receptor complex and because the question of drug-cell binding is left open by intention. Surprisingly Ahlquist did not notice Dale's hostile remarks on the receptor concepts and their supporters in the 1920s and 1930s. It had been already pointed out by Sir James Black that Dale neglected the results of receptor research and that this had hampered the consideration of Ahlquist's work, but – as already described in Chapter 4 – it was far more than only negligence. See Black (1988/89, pp. 60–9).

7 The Emergence of Molecular Pharmacology

1. Cook (1926/27).
2. Del Castillo and Katz (1955). See also Chapter 5 above.
3. Danielli and Davson (1935).
4. Cuello et al. (1988).
5. De Robertis et al. (1962) and De Robertis (1975).
6. See, for example, Azcurra and De Robertis (1967) and Schleifer and Eldefrawi (1974).
7. De Robertis (1975).
8. See, for example, Eldefrawi and Eldefrawi (1972).
9. De Robertis (1975).
10. Whittaker (1998).
11. For review see De Robertis (1975).
12. Whittaker (1998).
13. Chang and Lee (1963), Lee and Chang (1966) and Lee et al. (1967).
14. Hawgood (2002).
15. Changeux et al. (1970).
16. See for example, Lee et al. (1967).
17. Changeux et al. (1970).
18. Miledi et al. (1971).
19. Cited in De Robertis (1975); see also La Torre et al. (1970).
20. Cartaud et al. (1973).
21. Hucho and Changeux (1973).
22. See, for example, Miledi (1966).
23. Miledi (1973).
24. Katz and Miledi (1970).
25. Olsen et al. (1972).
26. Changeux (1971), De Robertis (1975) and Karlin (1974).
27. Miledi et al. (1971).
28. See Reynolds and Karlin (1978).
29. Katz and Miledi (1970).
30. Katz and Miledi (1970; 1971; 1973).
31. Colquhoun and Sakmann (1998).
32. See, for example, Cartaud et al. (1973).
33. For a review of knowledge of this time, see Conti-Tronconi and Raftery (1982).
34. Raftery et al. (1980).
35. Noda et al. (1982; 1983a; 1983b).

36. Mishina et al. (1985).
37. Ballivet et al. (1982).
38. Claudio et al. (1983).
39. See, for example, Toyoshima and Unwin (1988).
40. Kubalek et al. (1987).
41. See, for example, Schofield et al. (1987) and Levitan et al. (1988).
42. Ashcroft (2006).
43. See, for example, Reynolds (2007).

Conclusions

1. G.J.V. Nossal, Introduction, in Silverstein (2002, pp. xi–iii, here p. xi).
2. See, for example, Valenstein (2005).
3. Golinski (1998).
4. Hannaway and La Berge (1998).
5. Ackerknecht (1953) and Prüll (2003a).
6. This division between the two branches of medicine shapes for example the analysis of nineteenth-century medicine by William Bynum (1994).
7. As an example of studies supporting the alienation theory, see Howell (1995).
8. See for example the description by Thomas Schlich (1998).
9. Harwood (2000).
10. See Szöllösi-Janze (1998).
11. Goodman (2000, p. 143).
12. The term 'boundary idea' is used here basically in the same sense as 'boundary objects' in order to describe the usage of an idea with slightly different meanings in different scientific communities and to show the idea's potential to promote discussion among members of different scientific groups. For 'boundary objects', see above all Star and Griesemer (1989).
13. Lupton (1994).

Archival Sources

Rockefeller Archive Center (RAC), New York

Zettel Buch I, 1900, February 1 to 1900, December 26, folder 1, 2, 3, box 7, series II, (3) – 1, 2, 4 (No. of transcripts & index copy), Paul Ehrlich Collection, Rockefeller University Archives.

Zettel Buch II und Carcinom, 1900, December 25 to 1901, September 21, folder 1, 2, 3, box 8, series II, (3) – 1, 2, 4 (No. of transcripts & index copy), Paul Ehrlich Collection, Rockefeller University Archives.

Copir Buch II. Direktor, 1898, November 11 to 1899, February 21, folder 1, 2, 3, 4, box 4, series I, (3) – 1, 2, 4 (No. of transcripts & index copy), Paul Ehrlich Collection, Rockefeller University Archives.

Copir Buch III. Ehrlich, 1899, February 21 to 1899, July 31, folder 1, 2, 3, 4, box 5, series I, (3) – 1, 2, 4 (No. of transcripts & index copy), Paul Ehrlich Collection, Rockefeller University Archives.

Direktor, Ehrlich IX, 1902, September 25 to 1902, December 23, folder 1, 2, 3, 4, 5, 6, box 21, series V, (3-complete) – 1, 2, 4 (2-incomplete) (No. of transcripts & index copy), Paul Ehrlich Collection, Rockefeller University Archives.

Direktor, Ehrlich X, 1902, December 23 to 1903, January 26, folder 1, 2, 3, 4, 5, 6, box 22, series V, (3-complete) – 1, 2, 4 (2-incomplete) (No. of transcripts & index copy), Paul Ehrlich Collection, Rockefeller University Archives.

Direktor, Ehrlich XI, 1903, March 13 to 1903, July 17, box 23, series V, (mostly typescripts, no transcripts) (No. of Transcripts & index copy), Paul Ehrlich Collection, Rockefeller University Archives.

Direktor, Ehrlich XIII, 1903, December 21 to 1904, June 13, box 23, series V, (mostly typescripts, no transcripts) (No. of Transcripts & index copy), Paul Ehrlich Collection, Rockefeller University Archives.

Archive of the Humboldt-University, Berlin

Acta der Friedrich-Wilhelms-Universität Berlin. Habilitationen von 1880–1889, Medizinische Fakultät – Dekanat – , No.1342/1.

Acta betr. die Anstellung des Geheimen Medicinal Raths und Professors Dr. Gerhardt als dirigirender Arzt und Director der 2. medicinischen Universitäts-klinik, 1885, Kgl. Charité-Direction, No.437.

Acta betr. das Institut für Infectionskrankheiten, Kgl. Charité-Direction., No.2205, 1893–1895.

State Archive of Prussian Cultural Heritage (*Geheimes Staatsarchiv Preußischer Kulturbesitz* [GStA PK]), Berlin

Acta betr. die Einrichtung und die Verwaltung des (staatlichen) Institutes für Infektionskrankheiten in Berlin, vom Januar 1892 bis Dezember 1898, GStA PK I HA Rep.76 Kultusministerium, VIII B, No.2893.

Acta betr. das Institut für experimentelle Therapie zu Frankfurt a.m., vom Februar 1895 bis Dezember 1900, GStA PK I HA Rep.76 Kultusministerium, Vc Sekt.1, Tit.XI, Teil II, No.18, vol.1.
Nachlass Althoff B, GStA PK VI HA, Rep.92, No.33.
Nachlass Althoff A I, GStA PK VI HA, Rep.92, No.258.

University Archive of the University of Freiburg (UAF)

B 24 / 3743, Straub, Walther, von Augsburg (Personalakte) (no pagination).
B1 / 1225, (Lehrstuhl) Generalia. Medizinische Fakultät. Diener und Dienste. Den Lehrstuhl für Pharmakologie betr. (1907–1927).

Contemporary Medical Archive Centre (CMAC), London

Bayliss, Sir William, FRS (1860–1924), GC / 223 / A2.
British Pharmacological Society. General Minutes, Box 1, Vol. I, 1931–1935, SA / BPS / A 1, A 3, A 4.
Cushny, Professor Arthur Robertson, FRS (1866–1926), PP / ARC, A 4, D 1.
Dale, Professor Henry Hallett, OM, GBE, FRS (1875–1968), PP / HHD / 2, 15.
Dixon, Walter Ernest, FRS, FRCP, OBE, MD (1870–1931) and Myers (George) Norman, FRCP, MD, PhD (1898–1981), GC / 155 / A 1, B 4.
Elliott, Professor Thomas Renton, CBE, DSO, FRCP, FRS (1877–1961), GC / 42, 13, 15.
Gaddum, Sir John, FRS (1900–1965), GC / 213 / A.1.

University Archive of the University of Graz

Univ. Prof. Dr. Otto Loewi, geb. 3.6.1873. Personalakte.
Material Collection Otto Loewi.

Medical College of Georgia, Augusta, USA

Material in courtesy of Lois T. Ellison, Medical Historian in Residence, Provost Emeritus, Professor Emeritus of Medicine and Surgery.
Material in courtesy of Lowell Greenbaum, Emeritus Professor of Pharmacology and former Head of the Department of Pharmacology.
Material in courtesy of Richard E. White, Associate Professor of the Department of Pharmacology and Toxicology.

Department of Pharmacology and Toxicology, Medical College of Georgia, Augusta, USA

Ahlquist, Raymond P., Sympathomimetic Amines. Book I. Compound and Reference Numbers. R. P. Ahlquist 1945 (Laboratory Notebook).
Ahlquist, Raymond P., Sympathomimetic Amines. Book II (Laboratory Notebook).
Ahlquist, Raymond P., Folder (no signature).

Robert B. Greenblatt Library, Medical College of Georgia, Augusta, USA

Ahlquist, Raymond P., Folder (no signature).

Department of Manuscripts and University Archives, Cambridge, UK

Readers and Lecturers from 1878 to 1929, vol. V, Physiology – Talmudic, UA CUR 113.5.
Election of Professorships under Statute B. Chapter IX, Statutes (1926), Chapter XIV, UA CUR O.XIV.54.

Department of Physiology, University of Cambridge, UK

Box II: History of Lab Building 1912–14, Letter Accounts, to Prof. Langley, Deed of J. Physiol. from Mrs. Langley. Papers in connection with building of Physiol. Lab.

Bibliography

Abel, John J. (1926), 'Arthur Robinson Cushny and Pharmacology', *Journal of Pharmacology and Experimental Therapeutics*, 24: 265–86.

Adam, Matthias (2005), 'Integrating Research and Development: the Emergence of Rational Drug Design in the Pharmaceutical Industry', *Studies in History and Philosophy of Biological and Biomedical Sciences*, 36: 513–37.

Ackerknecht, Erwin H. (1953), *Rudolf Virchow: Doctor, Statesman, Anthropologist* (Madison, WI: University of Wisconsin Press).

Ackerknecht, Erwin H. (1974), 'The History of the Discovery of the Vegetative (Autonomic) Nervous System', *Medical History*, 18: 1–8.

Ahlquist, Raymond P. (1948), 'A Study of the Adrenotropic Receptors', *American Journal of Physiology*, 153: 586–600.

Ahlquist, Raymond P. (1958), 'Adrenergic Drugs', in Victor A. Drill (ed.), *Pharmacology in Medicine: a Collaborative Textbook*, 2nd edn (New York: McGraw-Hill), pp. 378–407.

Ahlquist, Raymond P. (1959), 'The Receptors for Epinephrine and Norepinephrine', *Pharmacological Review*, 11: 441–2.

Ahlquist, Raymond P. (1962), 'The Adrenotropic Receptor-Detector', *Archives Internationales de Pharmacodynamie et de Thérapie*, 139: 38–41.

Ahlquist, Raymond P. (1967), 'Development of the Concept of *Alpha* and *Beta* Adrenotropic Receptors', *Annals of the New York Academy of Sciences*, 139: 549–52.

Ahlquist, Raymond P. (1968), 'Agents which Block Adrenergic ß-Receptors', *Annual Review of Pharmacology and Toxicology*, 8: 259–72.

Ahlquist, Raymond P. (1973), 'Adrenergic Receptors: a Personal and Practical View', *Perspectives in Biology and Medicine*, 17: 119–22.

Ahlquist, Raymond P. (1976), 'Present State of Alpha and Beta Adrenergic Drugs. II: the Adrenergic Blocking Agents', *American Heart Journal*, 92: 804–7.

Ahlquist, Raymond P. (1980), 'Historical Perspective: Classification of Adrenoreceptors', *Journal of Autonomic Pharmacology*, 1: 101–6.

Ahlquist, Raymond P. and Bernhard Levy (1959), 'Adrenergic Receptive Mechanism of Canine Ileum', *Journal of Pharmacology and Experimental Therapeutics*, 127: 146–9.

Ahlquist, Raymond P. and R.A. Woodbury (1947), 'Influence of Drugs, and Uterine Activity upon Uterine Blood Flow', *Federation Proceedings*, 6: 305.

Ahlquist, Raymond P., Russell A. Huggins and R.A. Woodbury (1947), 'The Pharmacology of Benzyl-Imidazoline (Priscol)', *Journal of Pharmacology and Experimental Therapeutics*, 89: 271–88.

Alsberg, C. L. (1921), 'Chemical Structure and Physiological Action', *Journal of the Washington Academy of Sciences*, 11: 321–41.

Apolant, Hugo et al. (1914), *Paul Ehrlich: Eine Darstellung seines wissenschaftlichen Wirkens*, Festschrift zum 60. Geburtstage des Forschers (14. März 1914) (Jena: Gustav Fischer).

Arendt, Hans-Jürgen (1999), *Gustav Theodor Fechner: Ein deutscher Naturwissenschaftler und Philosoph im 19. Jahrhundert*, Daedalus, Europäisches Denken in deutscher Philosophie, Vol. 12 (Frankfurt/M. and Berlin: Peter Lang).

Arendt, Hans-Jürgen (2001), 'Gustav Theodor Fechner (1801–1887) und die Leipziger bürgerliche Gesellschaft im 19. Jahrhundert', *Naturwissenschaft. Technik. Medizin*, 9: 2–14.

Ariëns, E.J. (1954), 'Affinity and Intrinsic Activity in the Theory of Competitive Inhibition', *Archives Internationales de Pharmacodynamie et de Thérapie*, 99: 32–49.

Ariëns, E.J. (ed.) (1964), *Molecular Pharmacology: the Mode of Action of Biologically Active Compounds*, 2 vols (New York: Academic Press).

Arrhenius, Svante (1915), *Quantitative Laws in Biological Chemistry* (London: G. Bell and Sons Ltd.)

Arrhenius, Svante and Thorvald Madsen (1903), 'Anwendung der physikalischen Chemie auf das Studium der Toxine und Antitoxine', *Zeitschrift für Physikalische Chemie*, 44: 7–62.

Aschoff, Ludwig (1902), *Ehrlich's Seitenkettentheorie und ihre Anwendung auf die künstlichen Immunisierungsprozesse* (Jena: Gustav Fischer).

Ashcroft, Frances M. (2006), 'From Molecule to Malady', *Nature*, 440: 440–7.

Azcurra, J.M. and E. De Robertis (1967), 'Binding of dimethyl-C^{14}-d-tubocurarine, methyl-C^{14}-hexamethonium, and H^3-alloferine by Isolated Synaptic Membranes of Brain Cortex', *International Journal of Neuropharmacology*, 6 (1): 15–26.

Back, Marie-Luise (1986), 'Die Entwicklung des Freiburger Pharmakologischen Instituts 1907–1923', Med. Thesis, Freiburg i. Brsg.

Bacq, Z.M. (1983), 'Chemical Transmission of Nerve Impulses', in M.J. Parnham and J. Bruinvels (eds), *Discoveries in Pharmacology, Vol. I, Psycho- and Neuropharmacology* (Amsterdam: Elsevier Science Publishing), pp. 49–103.

Ballivet, M., J. Patrick, J. Lee and S. Heinemann (1982), 'Molecular Cloning of cDNA Coding for the γ Subunit of *Torpedo* Acetylcholine Receptor', *Proceedings of the National Academy of Sciences*, 79: 4466–70.

Barcroft, J. and A.V. Hill (1910), 'The Nature of Oxyhaemoglobin, with a Note on its Molecular Weight', *Journal of Physiology*, 39: 411–28.

Barger, G. and H.H. Dale (1910), 'Chemical Structure and Sympathomimetic Action of Amines', *Journal of Physiology*, 41: 19–59.

Bartlett, B. (2000), 'Health Issues: US Spending on Prescription Drugs', National Center for Policy Analysis', 4 September, http://www.who.int/medicines/publications/essentialmedicines/en/.

Bäumler, Ernst (1997), *Paul Ehrlich. Forscher für das Leben*, 3rd edn (Frankfurt/M.: Wötzel).

Bayliss, William Maddock (1915), *Principles of General Physiology* (London: Longmans, Green, and Co.)

Benison, Saul (1971), 'Oral History: a Personal View', in Edwin Clarke (ed.), *Modern Methods in the History of Medicine* (London: Athlone), pp. 286–305.

Bennett, Max R. (2000), 'The Concept of Transmitter Receptors: 100 Years On', *Neuropharmacology*, 39: 523–46.

Bericht über die Thätigkeit des Kgl. Instituts für Serumforschung und Serumprüfung zu Steglitz (1899), Juni 1896–September 1899. Zur Einweihung des Königl. Instituts für experimentelle Therapie Frankfurt/M. (Jena: Fischer).

Bickel, Marcel H. (2000), *Die Entwicklung zur experimentellen Pharmakologie 1790–1850. Wegbereiter von Rudolf Buchheim* (Basel: Schwabe & Co).

Binz, Carl (1890), 'Zur Geschichte der Pharmakologie in Deutschland', *Klinisches Jahrbuch*, 2: 3–74.

Black, James W. (1976), 'Ahlquist and the Development of Beta-Adrenoceptor Antagonists', *Postgraduate Medical Journal*, 52, Suppl. 4: 11–13.

Black, James W. (1988/89), 'Drugs, Hormones and Receptors: a Personal Reflection', *Hunterian Society Transactions*, 47: 60–9.

Black, James W. (1993), 'Drugs from Emasculated Hormones: the Principles of Synaptic Antagonism', Nobel Lecture, 8 December 1988, in Tore Frängsmyr and Jan Lindsten (eds), *Nobel Lectures, Physiology or Medicine 1981–1990* (Singapore: World Scientific Publishing Co.), pp. 418–40.

Black, James W. (2004), 'Interview with Sir James Black: Learning by Doing', *Molecular Interventions*, 4: 139–42.

Black, James W. and John S. Stephenson (1962), 'Pharmacology of a New Adrenergic Beta-Receptor-Blocking Compound (Nethalide)', *Lancet*, II: 311–14.

Black, James W., A.F. Crowther, R.G. Shanks, L.H. Smith and A.C. Dornhorst (1964), 'A New Adrenergic Beta-Receptor Antagonist', *Lancet*, I: 1080–1.

Boehm, Rudolf (1895), 'Einige Beobachtungen über die Nervenendwirkung des Curarin', *Archiv für experimentelle Pathologie und Pharmakologie*, 35: 16–22.

Bon, F., E. Lebrun, J. Gomel, R. Van Rapenbusch, J. Cartaud, J.L. Popot, and J.P. Changeux (1984), 'Image Analysis of the Heavy Form of the Acetylcholine Receptor from *Torpedo marmorata*', *Journal of Molecular Biology*, 176: 205–37.

Bordet, Jules (1899a), 'Le Mécanism de L'Agglutination', *Annales de l'Institut Pasteur*, 13: 225–50.

Bordet, Jules (1899b), 'Agglutination et Dissolution des Globules Rouges par le Sérum', *Annales de l'Institut Pasteur*, 13: 273–97.

Bordet, Jules (1900), 'Les Sérums Hémolytiques, Leurs Antitoxines et les Théories des Sérums Cytolytiques', *Annales de l'Institut Pasteur*, 14: 257–96.

Boruttau, Heinrich (1899), 'Erfahrungen über die Nebennieren', *Archiv für die gesamte Physiologie*, 78: 97–128.

Bowman, William C. (1999), 'Pharmacology: Past, Present and Future', *British Pharmacological Society Bulletin*, Winter: 9–12.

Bowman, W.C. and M.J. Rand (1980), *Textbook of Pharmacology*, 2nd edn (Oxford: Blackwell Scientific Publications).

Breathnach, C. S. (1978), 'A Century of Humours and Rumours', *Journal of the Irish College of Physicians and Surgeons*, 7(3): n.p.

Brock, A.J. (trans.) (1928), *Galen on the Natural Faculties* (London: William Heinemann).

Brocke, Bernhard vom (1980), 'Hochschul- und Wissenschaftspolitik in Preußen und im Deutschen Kaiserreich 1882–1907: das "System Althoff"', in Peter Baumgart (ed.), *Bildungspolitik in Preußen zur Zeit des Kaiserreichs* (Preußen in der Geschichte, 1) (Stuttgart: Klett-Cotta), pp. 9–118.

Brocke, Bernhard vom (ed.) (1991), *Wissenschaftsgeschichte und Wissenschaftspolitik im Industriezeitalter. Das 'System Althoff' in historischer Perspektive* (Hildesheim: Lax).

Brodie, T.G. and W.E. Dixon (1904), 'Contributions to the Physiology of the Lungs. Part II: On the Innervation of the Pulmonary Blood Vessels; and

Some Observations on the Action of Suprarenal Extract', *Journal of Physiology*, 30: 476–502.

Brunton, T. Lauder (1893), *Handbuch der Allgemeinen Pharmakologie und Therapie*, 3rd edn (Leipzig: F.A. Brockhaus).

Bruppacher-Cellier, Marianne (1971), 'Rudolf Buchheim (1820–1879) und die Entwicklung einer experimentellen Pharmakologie', Med. Thesis, Zurich.

Buchheim, Rudolf (1876), 'Ueber die Aufgaben und die Stellung der Pharmakologie an den deutschen Hochschulen', *Archiv für experimentelle Pathologie und Pharmakologie*, 5: 261–78.

Bulloch, William (1960), *The History of Bacteriology*, 2nd edn (London: Oxford University Press).

Burn, J.H. (1942), 'A.J. Clark 1885–1941', *Journal of Pharmacology and Experimental Therapeutics*, 75: 187–90.

Bynum, William F. (1970), 'Chemical Structure and Pharmacological Action: a Chapter in the History of 19th Century Molecular Pharmacology', *Bulletin of the History of Medicine*, 44: 518–38.

Bynum, William F. (1981a), *An Early History of the British Pharmacological Society* (published by the Society on the occasion of its 50th anniversary) (London: The British Pharmacological Society).

Bynum, William F. (1981b), 'Dale, Sir Henry Hallett', in *Dictionary of Scientific Biography*, Vol. 15, Suppl. 1 (New York: Charles Scribner's Sons), pp. 104–7.

Bynum, William F. (1994), *Science and the Practice of Medicine in the Nineteenth Century* (Cambridge: Cambridge University Press).

Bynum, William F. and Roy Porter (eds) (1988), *Brunonianism in Britain and Europe* (London: Wellcome Institute for the History of Medicine).

Cambrosio, Alberto, Daniel Jacobi and Peter Keating (1993), 'Ehrlich's "Beautiful Pictures" and the Controversial Beginnings of Immunological Imagery', *Isis*, 84: 662–99.

Cannon, Walter Bradford and Arturo Rosenblueth (1933), 'Studies on Conditions of Activity in Endocrine Organs, XXIX. Sympathin E and Sympathin I', *American Journal of Physiology*, 104: 557–74.

Cannon, Walter Bradford and Arturo Rosenblueth (1937), *Autonomic Neuro-Effector Systems* (New York: The Macmillan Company).

Carrier, Gerald O. (1986), 'Evolution of the Dual Adrenergic Receptor Concept: Key to Past Mysteries and Modern Therapy', in M.J. Parnham and J. Bruinvels (eds), *Discoveries in Pharmacology, Vol. 3, Pharmacological Methods, Receptors & Chemotherapy* (Amsterdam: Elsevier Science Publishers), pp. 203–21.

Carrier, Gerald O. and Marshal Shlafer (1984), 'Raymond P. Ahlquist. 26 July 1914–15 April 1983', *Trends in Pharmacological Sciences*, 5: 41–4.

Cartaud, J., E.L. Benedetti, J.B. Cohen, J.C. Meunier and J.P. Changeux (1973), 'Presence of a Lattice Structure in Membrane Fragments Rich in Nicotinic Receptor Protein from the Electric Organ of *Torpedo marmorata*', *FEBS Letters*, 33: 109–13.

Chagas, C., E. Penna-Franca, K. Nishie and R.J. Gargia (1958), 'A study of the Specificity of the Complex Formed by Gallamine Triethiodide with a Macromolecular Constituent of the Electric Organ', *Archives of Biochemistry and Biophysics*, 75(1): 251–9.

Chalmers, Alan F. (1986), *Wege der Wissenschaft. Einführung in die Wissenschaftstheorie* (Berlin, Heidelberg and New York: Springer) (5th edn, 2001).

Chang, C.C. and C.Y. Lee (1963), 'Isolation of Neurotoxins from the Venom of *Bungarus Multicinctus* and their Modes of Neuromuscular Blocking Action', *Archives Internationales de Pharmacodynamie et de Thérapie*, 144: 241–57.

Changeux, J.P. (1971), 'Studies on the Cholinergic Receptor Protein of *Electrophorus electricus*, I. An Assay *in vitro* for the Cholinergic Receptor Site and Solubilization of the Receptor Protein from Electric Tissue', *Molecular Pharmacology*, 7: 538–53.

Changeux, J.P., M. Kasai and C.Y. Lee (1970a), 'Use of Snake Venom Toxin to Characterize the Cholinergic Receptor Protein', *Proceedings of the National Academy of Sciences*, 67: 1241–7.

Changeux, J.P., J.P. Kasai, M. Huchet and J.C. Meunier (1970b), 'Extraction from Electric Tissue of Electrophorus of a Protein Presenting Several Typical Properties Characteristic of the Physiological Receptor of Acetylcholine', *C. R. Acad. Sci. Hebd. Seances Acad. Sci.*, D 270: 2864–7.

Church, R. and T. Tansey (2007), *Burroughs Wellcome & Co: Knowledge, Trust, Profit and the Transformation of the British Pharmaceutical Industry, 1880–1940* (Lancaster: Crucible Books).

Clark, Alfred Joseph (1912), 'The Influence of Ions upon the Action of Digitalis', *Proceedings of the Royal Society of Medicine*, 5: 181–97.

Clark, A.J. (1913a), 'The Mode of Action of Strophanthin upon Cardiac Tissue', *Journal of Pharmacology and Experimental Therapeutics*, 5: 215–34.

Clark, A.J. (1913b), 'The Action of Ions and Lipoids upon the Frog's Heart', *Journal of Physiology*, 47: 66–107.

Clark, A.J. (1922), 'The Mode of Action of Potassium Ions upon Isolated Organs', *Journal of Pharmacology and Experimental Therapeutics*, 18: 423–47.

Clark, A.J. (1926a), 'The Reaction between Acetyl Choline and Muscle Cells', *Journal of Physiology*, 61: 530–46.

Clark, A.J. (1926b), 'The Antagonism of Acetyl Choline by Atropine', *Journal of Physiology*, 61: 547–56.

Clark, A.J. (1926c), 'The Mode of Action of Potassium Chloride on Muscles', *Journal of Pharmacology and Experimental Therapeutics*, 29: 311–24.

Clark, A.J. (1927a), *Comparative Physiology of the Heart* (Cambridge: Cambridge University Press).

Clark, A.J. (1927b), 'The Reaction between Acetyl Choline and Muscle Cells. Part II', *Journal of Physiology*, 64: 123–43.

Clark, A.J. (1929), *Applied Pharmacology*, 3rd edn (London: J. & A. Churchill).

Clark, A.J. (1932), 'Walter Ernest Dixon – 1871–1931', *Proceedings of the Royal Society of London*, series B, 110: xxix–xxxi.

Clark, A.J. (1933), *The Mode of Action of Drugs on Cells* (London: Edward Arnold & Co.)

Clark, A.J. (1937a), *General Pharmacology* (*Handbuch der Experimentellen Pharmakologie*, founded by A. Heffter, *Ergänzungswerk*, edited by W. Heubner and J. Schüller, vol. 4) (Berlin: Verlag von Julius Springer).

Clark, A.J. (1937b), 'Discussion on the Chemical and Physical Basis of Pharmacological Action', 12 November 1936, Opening Address, *Proceedings of the Royal Society of London*, series B, 121: 580–4.

Clark, A.J. (1938), *Patent Medicines* (*Fact*, No. 14, London).

Clark, A.J. (1940), Obituary note on George Barger, in *Proceedings of the Royal Society of Edinburgh*, Vol. LIX (Edinburgh: Neill and Company), pp. 260–1.

Clark, A.J. and J. Raventós (1937), 'The Antagonism of Acetylcholine and of Quaternary Ammonium Salts', *Quarterly Journal of Experimental Physiology*, 26: 375–92.

Clark, A.J., M.G. Eggleton, P. Eggleton, R. Gaddie and C.P. Stewart (1938), *The Metabolism of the Frog's Heart* (Edinburgh and London: Oliver and Boyd).

Clark, George and Frederick H. Kasten (1983), *History of Staining* (Baltimore: Williams and Wilkins) (revised edn of Harold Joel Conn et al., *The History of Staining*, 2nd edn, 1948).

Clarke, E. and L.S. Jacyna (1987), *Nineteenth-Century Origins of Neuroscientific Concepts* (Berkeley: University of California Press).

Classen, M., F.H. Franken and D. Gericke (1995), 'Friedrich Theodor Frerichs in Berlin', *Deutsche Medizinische Wochenschrift*, 120: 1334–7.

Claudio, T., M. Ballivet, J. Patrick and S. Heinemann (1983), 'Nucleotide and Deduced Amino Acid Sequences of *Torpedo californica* Acetylcholine Receptor γ Subunit', *Proceedings of the National Academy of Sciences*, 80: 1111–15.

Colquhoun, D. (2006), 'The Quantitative Analysis of Drug-Receptor Interactions: a Short History', *Trends in Pharmacological Sciences*, 27: 149–57.

Colquhoun, D. and B. Sakmann (1998), 'From Muscle Endplate to Brain Synapses: a Short History of Synapses and Agonist-Activated Ion Channels', *Neuron*, 20: 381–7.

Conti-Tronconi, B.M. and M.A. Raftery (1982), 'The Nicotinic Cholinergic Receptor: Correlation of Molecular Structure with Functional Properties', *Annual Review of Biochemistry*, 51: 491–530.

Cook, R.P. (1926/27), 'Antagonism of Acetylcholine by Methylene Blue', *Journal of Physiology*, 62: 160–5.

Cooter, Roger and John Pickstone (2000), 'Introduction', in Roger Cooter and John Pickstone (eds), *Medicine in the 20th Century* (Amsterdam: Harwood Academic Publishers), pp. xiii–xix.

Cozzens, E. (1989), *Social Control and Multiple Discovery in Science: the Opiate Receptor Case* (Albany: State University of New York Press).

Craig, Stewart P. (1988), 'If Alive Today, MCG Professor Might Have Won Nobel', *Augusta Chronicle*, 22 October.

Cramer, Friedrich (1997), 'Emil Fischers Schlüssel-Schloß-Hypothese der Enzymwirkung – 100 Jahre danach', in Hans-Jörg Rheinberger, Michael Hagner and Bettina Wahrig-Schmidt (eds), *Räume des Wissens. Repräsentation, Codierung, Spur* (Berlin: Akademie-Verlag), pp. 191–212.

Crist, Eileen and Alfred I. Tauber (1997), 'Debating Humoral Immunity and Epistemology: the Rivalry of the Immunochemists Jules Bordet and Paul Ehrlich', *Journal of the History of Biology*, 30: 321–56.

Cuello, A.A., A. Pellegrino de Iraldi and J.P. Saavedra (1988), 'In Memoriam: Eduardo De Robertis 1913–1988', *Journal of Neurochemistry*, 51(6): 1964–5.

Cushny, Arthur Robertson (1901), *A Text-book of Pharmacology and Therapeutics or the Action of Drugs in Health and Disease*, 2nd edn (London: Rebman Limited).

Cushny, A.R. (1903), *A Text-book of Pharmacology and Therapeutics or the Action of Drugs in Health and Disease*, 3rd edn (London: Rebman Limited).

Cushny, A.R. (1907), *A Text-book of Pharmacology and Therapeutics or the Action of Drugs in Health and Disease*, 4th edn (London: Rebman Limited).

Cushny, A.R. (1908), 'The Action of Optical Isomers. III. Adrenalin', *Journal of Physiology*, 37: 130–8.

Cushny, A.R. (1910a), *A Text-book of Pharmacology and Therapeutics or the Action of Drugs in Health and Disease*, 5th edn (London: J. & A. Churchill).

Cushny, A.R. (1910b), 'The Action of Atropine, Pilocarpine and Physostigmine', *Journal of Physiology*, 41: 233–45.

Cushny, A.R. (1915a), 'Quantitative Observations on Antagonism', *Journal of Pharmacology and Experimental Therapeutics*, 6: 439–50.

Cushny, A.R. (1915b), *A Text-book of Pharmacology and Therapeutics or the Action of Drugs in Health and Disease*, 6th edn (London: J. & A. Churchill).

Cushny, A.R. (1918), *A Text-book of Pharmacology and Therapeutics or the Action of Drugs in Health and Disease*, 7th edn (London: J. & A. Churchill).

Cushny, A.R. (1924), *A Text-book of Pharmacology and Therapeutics or the Action of Drugs in Health and Disease*, 8th edn (Philadelphia and New York: Lea & Febiger).

Cushny, Arthur Robertson, Obituary (1926a), *The British Medical Journal*, 6 March, pp. 455–7.

Cushny, Arthur Robertson, Obituary (1926b), *Lancet*, 6 March, pp. 519–20.

Cushny, A.R. (1928), *A Text-book of Pharmacology and Therapeutics or the Action of Drugs in Health and Disease*, 9th edn, thoroughly revised by C.W. Edmunds and J.A. Gunn (London: J. & A. Churchill).

Cushny, A.R. (1934), *A Text-book of Pharmacology and Therapeutics or the Action of Drugs in Health and Disease*, 10th edn, thoroughly revised by C.W. Edmunds and J.A. Gunn (London: J. & A.Churchill).

Dale, Henry Hallett (1899/1900), 'On some Numerical Comparisons of the Centripetal and Centrifugal Medullated Nerve-Fibres Arising in the Spinal Ganglia of the Mammal', *Journal of Physiology*, 25: 196–206.

Dale, H.H. (1901/02), 'Observations, Chiefly by the Degeneration Method, on Possible Efferent Fibres in the Dorsal Nerve-Roots of the Toad and Frog', *Journal of Physiology*, 27: 350–5.

Dale, H.H. (1905/06), 'The Contractile Mechanism of the Gall-bladder and its Extrinsic Nervous Control, *Journal of Physiology*, 33: 138–55.

Dale, H.H. (1906), 'On some Physiological Actions of Ergot', *Journal of Physiology*, 34: 163–206.

Dale, H.H. (1920), 'Chemical Structure and Physiological Action', *Johns Hopkins Hospital Bulletin*, 31: 373–80.

Dale, H.H. (1921), 'Recent Tendencies in Chemotherapy', *Proceedings of the Royal Society of Medicine*, Therap. Section 14: 7–15.

Dale, H.H. (1923), 'Chemotherapy', *Physiological Review*, 3: 359–93.

Dale, H.H. (1926a), 'Arthur Robertson Cushny – 1866–1926', *Proceedings of the Royal Society*, B 100: xix–xxvii.

Dale, H.H. (1926b), 'Arthur Robertson Cushny', *Archives Internationales de Pharmacodynamie et de Thérapie*, 32: 1–8.

Dale, H.H. (1932), 'Some Therapeutic Problems of the Future', *Pharmaceutical Journal and Pharmacist*, 129: 515–17.

Dale, H.H. (1934), 'Chemical Ideas in Medicine and Biology', *Science*, 80: 343–9.

Dale, H.H. (1935a), 'Some Epochs in Medical Research', The Harveian Oration, *British Medical Journal*, II: 771–7.

Dale, H.H. (1935b), 'Pharmacology and Nerve-Endings', Walter Ernest Dixon Memorial Lecture, *Proceedings of the Royal Society of Medicine*, 28: 319–32.

Dale, H.H. (1941), 'George Barger: 1878–1939', in *Obituary Notices of Fellows of the Royal Society*, Vol. 3, 1939–1941 (London: Morrison & Gibb Ltd.), pp. 63–85.

Dale, H.H. (1943), 'Modes of Drug Action: a General Discussion', General Introductory Address, *Transactions of the Faraday Society*, 39: 319–22.

Dale, H.H. (1949), 'Introduction', in Martha Marquardt, *Paul Ehrlich* (London: Heinemann), pp. xiii–xx.

Dale, H.H. (1953), *Adventures in Physiology. With Excursions into Autopharmacology*, ed. by the Wellcome Trust (London: Pergamon).

Dale, H.H. (1954), 'Walter Bradford Cannon, 1871–1945', in *Obituary Notices of Fellows of the Royal Society*, Vol. V, 1945–1948 (London: Morrison & Gibb Ltd.), pp. 407–23.

Dale, H.H. (1956), 'Introduction', in Fred Himmelweit (ed.), *The Collected Papers of Paul Ehrlich*, Vol. I (London and New York: Pergamon), pp. 1–18.

Dale, H.H. (1961), 'Thomas Renton Elliott, 1877–1961', *Biographical Memoirs of Fellows of the Royal Society*, 7: 53–74.

Dale, H.H. (1965), 'Some Recent Extensions of the Chemical Transmission of the Effects of Nerve Impulses', Nobel Lecture, 12 December 1936, in *Nobel Lectures Including Presentation Speeches and Laureate's Biographies, Physiology or Medicine 1922–1941* (Amsterdam, London, New York: Elsevier, for the Nobel Foundation), pp. 402–13.

Dale, H.H. and H.S. Gasser (1926), 'The Pharmacology of Denervated Mammalian Muscle. Part I. The Nature of the Substances Producing Contracture', *Journal of Pharmacology*, 29: 53–61.

Dannenberg, H. (1930), 'Über Hefegärung. Zur Frage der Gültigkeit der Arndt-Schulzschen Regel', *Archiv für Experimentelle Pharmakologie*, 154: 211–21.

Daniel, Ute (2001), 'Einleitung: Kulturgeschichte – und was sie nicht ist', in *Kompendium Kulturgeschichte. Theorien, Praxis, Schlüsselwörter* (Frankfurt am Main: Suhrkamp), pp. 7–25.

Danielli, J.F. and H. Davson (1935), 'The Permeability of Thin Films', *Journal of Cellular and Comparative Physiology*, 5: 495–508.

De Burgh Daly, I. and A.J. Clark (1921), 'The Action of Ions upon the Frog's Heart', *Journal of Physiology*, 54: 367–83.

De Jongh, D.K. (1964), 'Some Introductory Remarks on the Conception of Receptors', in E.J. Ariëns (ed.), *Molecular Pharmacology: the Mode of Action of Biologically Active Compounds*, 2 vols (New York: Academic Press), vol. 1, pp. xiii–xvi.

De Robertis, E. (1975), *Synaptic Receptors: Isolation and Molecular Biology* (Modern Pharmacology-Toxicology 4) (New York: M. Dekker).

De Robertis, E. (1982), 'Comment on This Week's Citation Classic', *Current Contents*, 6 September: 22.

De Robertis, E. and H.S. Bennett (1955), 'Some Features of Submicroscopic Morphology of Synapses in the Frog and Earthworm', *Journal of Biophysical and Biochemical Cytology*, 9: 229–35.

De Robertis, E., G. Rodriguez de Lores Arnaiz and A. Pellegrino de Iraldi (1962), 'Isolation of Synaptic Vesicles from Nerve Endings of the Rat Brain', *Nature*, 194: 794–5.

Debru, A. (ed.) (1977), *Galen on Pharmacology. Philosophy, History and Medicine* (Leiden: Brill).

Dehmel, G. (1996), *Die Arzneimittel in der Physikotheologie* (Münster: LIT Verlag).

Del Castillo, J. and B. Katz (1955), 'On the Localization of Acetylcholine Receptors', *Journal of Physiology*, 128(1):157–81.

Deriso, Christine (1988), 'Ahlquist Laid Foundation for Nobel', in *Beeper* (The Medical College of Georgia), 10(5), 2 November: 1.

Dietrich, S. and O. Loewi (1927), 'Untersuchungen über Diabetes und Insulinwirkung. VII. Mitteilung', *Pflügers Archiv für die gesamte Physiologie des Menschen und der Tiere*, 215: 78–94.

Dietrich, S., H. Häusler and O. Loewi (1926), 'Weiteres über Insulinwirkung und Diabetes', *Klinische Wochenschrift*, 5: 414.

Dietrich, S., H. Häusler and O. Loewi (1927), 'Weiteres über den insulinantagonistischen Stoff', *Klinische Wochenschrift*, 6: 856.

Dixon, Walter Ernest (1906), *A Manual of Pharmacology* (London: Edward Arnold).

Dixon, Walter Ernest (1908), *A Manual of Pharmacology*, 2nd edn (London: Edward Arnold).

Dixon, Walter Ernest (1913), *A Manual of Pharmacology*, 3rd edn (London: Edward Arnold).

Dixon, Walter Ernest (1915), *A Manual of Pharmacology*, 4th edn (London: Edward Arnold).

Dixon, Walter Ernest (1921), *A Manual of Pharmacology*, 5th edn (London: Edward Arnold).

Dixon, Walter Ernest (1925), *A Manual of Pharmacology*, 6th edn (London: Edward Arnold).

Dixon, Walter Ernest (1926), 'Obituary, Arthur Robertson Cushny', *Lancet*, 6 March: 519–20.

Dixon, Walter Ernest (1929), *A Manual of Pharmacology*, 7th edn (London: Edward Arnold).

Dixon, Walter Ernest and P. Hamill (1909), 'The Mode of Action of Specific Substances with Special Reference to Secretin', *Journal of Physiology*, 38: 314–36.

Dolman, Claude E. (1968), 'Paul Ehrlich and William Bulloch: a Correspondence and Friendship (1896–1914)', *Clio Medica*, 3: 65–84.

Dolman, Claude E. (1981), 'Paul Ehrlich', in *Dictionary of Scientific Biography*, Vol. 3, 10th edn (New York: Charles Scribner's Sons), pp. 295–305.

Dönitz, Wilhelm (1899), 'Bericht über die Thätigkeit des Königl. Instituts für Serumforschung und Serumprüfung zu Steglitz, Juni 1896–September 1899', *Klinisches Jahrbuch*, 7: 359–84.

Döring, H.J. (1996), 'Das isoliert perfundierte Herz nach Langendorff – Geschichte und Gegenwart, Modifikationen und Applikationen', *European Surgery*, 28: 328–33.

'Dr. Raymond Ahlquist, Noted Researcher, Dies' (1983), *Beeper* (The Medical College of Georgia), 4(16), 20 April: 1.

Drews, J. (2002), 'Drug Discovery: a Historical Perspective', *Science*, 287(5460): 1960–5.

Du Bois-Reymond, R. (1926), 'John Newport Langley zum Gedächtnis', *Ergebnisse der Physiologie*, 25: xv–xix.

Dungern, E. Freiherr von (1914), 'Rezeptorenspezifität', in Hugo Apolant et al. (eds), *Paul Ehrlich* (Jena: Gustav Fischer), pp. 162–5.

Dunlop, D.M. (1940–42), 'The Late Professor A.J. Clark', *University of Edinburgh Journal*, 11: 165–6.

Dupont, J.-C. (1999), *Histoire de la Neurotransmission* (Paris: Presses Universitaires de France).

Earles, M.P. (1961), 'Studies in the Development of Experimental Pharmacology in the Eighteenth and Early Nineteenth Centuries', PhD thesis, University of London.

Eckart, Wolfgang U. (1991), 'Friedrich Althoff und die Medizin', in Bernhard vom Brocke (ed.), *Wissenschaftsgeschichte und Wissenschaftspolitik im Industriezeitalter. Das 'System Althoff' in historischer Perspektive* (Hildesheim: Lax), pp. 375–404.

Eckmann, Maria Luise (1959), 'Die Doktorarbeit Paul Ehrlichs und ihre Bedeutung für die Geschichte der histologischen Färbung', Thesis, University of Hamburg.

Ehrlich, Paul (1877), 'Beiträge zur Kenntnis der Anilinfärbungen und ihrer Verwendung in der mikroskopischen Technik' (Archiv für mikroskopische Anatomie), in Himmelweit (ed.), *The Collected Papers of Paul Ehrlich*, Vol. I, pp. 19–28.

Ehrlich, Paul (1878), 'Beiträge zur Theorie und Praxis der histologischen Färbung', Thesis, Leipzig, in Himmelweit (ed.), *The Collected Papers of Paul Ehrlich*, Vol. I, pp. 29–64 (English translation, pp. 65–98).

Ehrlich, Paul (1882a), 'Beiträge zur Ätiologie und Histologie pleuritischer Exsudate' (Charité-Annalen), in Himmelweit (ed.), *The Collected Papers of Paul Ehrlich*, Vol. I, pp. 290–310.

Ehrlich, Paul (1882b), 'Über einen Fall von Phosphorvergiftung mit symmetrischer Gangraena pedum' (Charité-Annalen), in Himmelweit (ed.), *The Collected Papers of Paul Ehrlich*, Vol. I, pp. 526–9.

Ehrlich, Paul (1882c), 'Über provocirte Fluorescenzerscheinungen am Auge' (Deutsche Medizinische Wochenschrift), in Himmelweit (ed.), *The Collected Papers of Paul Ehrlich*, Vol. I, pp. 344–53.

Ehrlich, Paul (1882d), 'Modification der von Koch angegebenen Methode der Färbung von Tuberkelbazillen' (Deutsche Medizinische Wochenschrift), in Himmelweit (ed.), *The Collected Papers of Paul Ehrlich*, Vol. I, pp. 311–13.

Ehrlich, Paul (1882e), 'Referat über die gegen R. Koch's Entdeckung der Tuberkelbacillen neuerlichst hervorgetretenen Einwände' (Deutsche Medizinische Wochenschrift), in Himmelweit (ed.), *The Collected Papers of Paul Ehrlich*, Vol. I, pp. 322–9.

Ehrlich, Paul (1883a), 'Sulfodiazobenzol, ein Reagenz auf Bilirubin' (Zentralblatt für klinische Medizin), in Himmelweit (ed.), *The Collected Papers of Paul Ehrlich*, Vol. I, pp. 630–1.

Ehrlich, Paul (1883b), 'Über die Sulfodiabenzol-Reaction' (Zentralblatt für klinischen Medizin), in Himmelweit (ed.), *The Collected Papers of Paul Ehrlich*, Vol. I, pp. 632–42.

Ehrlich, Paul (1883c), 'Über eine neue Harnprobe' (Charité-Annalen), in Himmelweit (ed.), *The Collected Papers of Paul Ehrlich*,Vol. I, pp. 619–29.

Ehrlich, Paul (1883d), 'Über das Vorkommen von Glykogen im diabetischen und im normalen Organismus' (Zeitschrift für klinische Medizin), in Himmelweit (ed.), *The Collected Papers of Paul Ehrlich*, Vol. I, pp. 103–12.

Ehrlich, Paul (1885), 'Das Sauerstoff-Bedürfniss des Organismus. Eine farbenanalytische Studie', Habilitation-thesis (Berlin: Hirschwald), in Himmelweit (ed.), *The Collected Papers of Paul Ehrlich*, Vol. I, pp. 364–432 (English translation, pp. 433–96).

Ehrlich, Paul (1886a), 'Nachträgliche Bemerkungen zur Diazoreaction' (Charité-Annalen), in Himmelweit (ed.), *The Collected Papers of Paul Ehrlich*, Vol. I, pp. 643–5.

Ehrlich, Paul (1886b), 'Experimentelles und Klinisches über Thallin' (Deutsche Medizinische Wochenschrift), in Himmelweit (ed.), *The Collected Papers of Paul Ehrlich*, Vol. I, pp. 542–51.

Ehrlich, Paul (1890), 'Studien in der Cocainreihe' (Deutsche Medizinische Wochenschrift), in Himmelweit (ed.), *The Collected Papers of Paul Ehrlich*, Vol. I, pp. 559–66.

Ehrlich, Paul (1891a), 'Über neuere Erfahrungen in der Behandlung der Tuberkulose nach Koch, insbesondere der Lungenschwindsucht' (Transactions from the 7th International Congress of Hygiene and Demography), in Himmelweit (ed.), *The Collected Papers of Paul Ehrlich*, Vol. II, pp. 13–20.

Ehrlich, Paul (1891b), 'Zur Geschichte der Granula', repr. from *Farbenanalytische Untersuchungen zur Histologie und Klinik des Blutes* (Berlin: Hirschwald), in Himmelweit (ed.), *The Collected Papers of Paul Ehrlich*, Vol. I, pp. 166–8.

Ehrlich, Paul (1891c), 'Experimentelle Untersuchungen über Immunität I. Über Ricin' (Deutsche Medizinische Wochenschrift), in Himmelweit (ed.), *The Collected Papers of Paul Ehrlich*, Vol. II, pp. 21–6.

Ehrlich, Paul (1891d), 'Experimentelle Untersuchungen über Immunität II. Über Abrin' (Deutsche Medizinische Wochenschrift), in Himmelweit (ed.), *The Collected Papers of Paul Ehrlich*, Vol. II, pp. 27–30.

Ehrlich, Paul (1892), 'Über Immunität durch Vererbung und Säugung' (Zeitschrift für Hygiene und Infektionskrankheiten), in Himmelweit (ed.), *The Collected Papers of Paul Ehrlich*, Vol. II, pp. 31–44.

Ehrlich, Paul (1894), 'Über Gewinnung, Werthbestimmung und Verwerthung des Diphtherieheilserums' (Hygienische Rundschau), in Himmelweit (ed.), *The Collected Papers of Paul Ehrlich*, Vol. II, pp. 80–3.

Ehrlich, Paul (1897a), 'Die Wertbemessung des Diphtherieheilserums und deren theoretische Grundlagen' (Klinisches Jahrbuch), in Himmelweit (ed.), *The Collected Papers of Paul Ehrlich*, Vol. II, pp. 86–106 (English translation, pp. 107–25).

Ehrlich, Paul (1897b), 'Zur Kenntnis der Antitoxinwirkung' (Fortschritte der Medizin), in Himmelweit (ed.), *The Collected Papers of Paul Ehrlich*, Vol. II, pp. 84–5.

Ehrlich, Paul (1898), 'Über die Constitution des Diphtheriegiftes' (Deutsche Medizinische Wochenschrift), in Himmelweit (ed.), *The Collected Papers of Paul Ehrlich*, Vol. II, pp. 126–33.

Ehrlich, Paul (1900a), 'Die Diazo- und Azomethin-Reactionen' (draft, probably written in 1900), in Himmelweit (ed.), *The Collected Papers of Paul Ehrlich*, Vol. I, pp. 646–50.

Ehrlich, Paul (1900b), 'On Immunity with Special Reference to Cell Life', Croonian Lecture (Proceedings of the Royal Society), in Himmelweit (ed.), *The Collected Papers of Paul Ehrlich*, Vol. II, pp. 178–95.

Ehrlich, Paul (1901a), 'Über die Dimethylamidobenzaldehydreaction' (Medizinische Woche), in Himmelweit (ed.), *The Collected Papers of Paul Ehrlich*, Vol. I, pp. 651–3.

Ehrlich, Paul (1901b), 'Die Schutzstoffe des Blutes' (Verhandlungen der 73. Versammlung der Gesellschaft der Naturforscher und Ärzte), in Himmelweit (ed.), *The Collected Papers of Paul Ehrlich*, Vol. II, p. 298–315.

Ehrlich, Paul (1902), 'Über die Beziehung von chemischer Constitution, Vertheilung und pharmakologischer Wirkung', from *Festschrift für Ernst v. Leyden* (Berlin: Hirschwald), in Himmelweit (ed.), *The Collected Papers of Paul Ehrlich*, Vol. I, pp. 570–95 (English translation, pp. 596–618).

Ehrlich, Paul (1903a), 'Toxin und Antitoxin. Entgegnung auf den neuesten Angriff Grubers' (Münchener Medizinische Wochenschrift), in Himmelweit (ed.), *The Collected Papers of Paul Ehrlich*, Vol. II, pp. 368–90.

Ehrlich, Paul (1903b), 'Toxin und Antitoxin. Entgegnungen auf Grubers Replik' (Münchener Medizinische Wochenschrift), in Himmelweit (ed.), *The Collected Papers of Paul Ehrlich*, Vol. II, pp. 391–4.

Ehrlich, Paul (1904), 'Über den Receptorenapparat der rothen Blutkörperchen', from *Gesammelte Arbeiten zur Immunitätsforschung* (Berlin: Hirschwald), in Himmelweit (ed.), *The Collected Papers of Paul Ehrlich*, Vol. II, p. 316–23.

Ehrlich, Paul (1907), 'Das Königliche Institut für experimentelle Therapie zu Frankfurt a.M.' (*Festschrift zum XIV. Internationalen Kongreß für Hygiene und Demographie, Berlin 1907*, dargeboten von dem Preußischen Minister der geistlichen, Unterrichts- und Medizinalangelegenheiten) (Jena: Fischer).

Ehrlich, Paul (1909), 'Über Partialfunktionen der Zelle' (Nobel Lecture), in Himmelweit (ed.), *The Collected Papers of Paul Ehrlich*, Vol. III, pp. 171–82.

Ehrlich, Paul and Paul Guttmann (1891a), 'Über die Wirkung von Methylenblau bei Malaria' (Berliner Klinische Wochenschrift), in Himmelweit (ed.), *The Collected Papers of Paul Ehrlich*, Vol. III, pp. 9–14.

Ehrlich, Paul and Paul Guttmann (1891b), 'Die Wirksamkeit kleiner Tuberkulindosen gegen Lungenschwindsucht' (Deutsche Medizinische Wochenschrift), in Himmelweit (ed.), *The Collected Papers of Paul Ehrlich*, Vol. II, pp. 7–12.

Ehrlich, Paul and Hermann Kossel (1894), 'Über die Anwendung des Diphtherieantitoxins' (Zeitschrift für Hygiene und Infektionskrankheiten), in Himmelweit (ed.), *The Collected Papers of Paul Ehrlich*, Vol. II, pp. 61–2.

Ehrlich, Paul and A. Lazarus (1900), *Histology of the Blood* (Cambridge: Cambridge University Press) (Revised translation of P. Ehrlich and A. Lazarus, 'Die Anaemie', in H. Nothnagel, *Specielle Pathologie und Therapie* (Vienna, Hölder, 1898)), in Himmelweit (ed.), *The Collected Papers of Paul Ehrlich*, Vol. I, pp. 181–268.

Ehrlich, Paul and A. Leppmann (1890), 'Über schmerzstillende Wirkung des Methylenblau' (Deutsche Medizinische Wochenschrift), in Himmelweit (ed.), *The Collected Papers of Paul Ehrlich*, Vol. I, pp. 555–8.

Ehrlich, Paul and Julius Morgenroth (1899a), 'Zur Theorie der Lysinwirkung' (Berliner Klinische Wochenschrift), in Himmelweit (ed.), *The Collected Papers of Paul Ehrlich*, Vol. II, pp. 143–9 (Engish translation, pp. 150–5).

Ehrlich, Paul and Julius Morgenroth (1899b), 'Über Hämolysine. Zweite Mittheilung' (Berliner Klinische Wochenschrift), in Himmelweit (ed.), *The Collected Papers of Paul Ehrlich*, Vol. II, pp. 156–64 (English translation, pp. 165–72).

Ehrlich, Paul and Julius Morgenroth (1900a), 'Über Haemolysine. Dritte Mittheilung' (Berliner Klinische Wochenschrift), in Himmelweit (ed.), *The Collected Papers of Paul Ehrlich*, Vol. II, pp. 196–204 (English translation, pp. 205–12).

Ehrlich, Paul and Julius Morgenroth (1900b), 'Über Hämolysine. Dritte Mittheilung' (Berliner Klinische Wochenschrift), in Himmelweit (ed.), *The Collected Papers of Paul Ehrlich*, Vol. II., pp. 196–204 (English translation, pp. 205–12).

Ehrlich, Paul and Julius Morgenroth (1900c), 'Über Hämolysine. Vierte Mittheilung' (Berliner Klinische Wochenschrift), in Himmelweit (ed.), *The Collected Papers of Paul Ehrlich*, Vol. II, pp. 213–23 (English translation, pp. 224–33).

Ehrlich, Paul and Julius Morgenroth (1901a), 'Über Hämolysine. Fünfte Mittheilung' (Berliner Klinische Wochenschrift), in Himmelweit (ed.), *The Collected Papers of Paul Ehrlich*, Vol. II, pp. 234–45 (English translation, pp. 246–55).

Ehrlich, Paul and Julius Morgenroth (1901b), 'Über Hämolysine. Sechste Mittheilung' (Berliner Klinische Wochenschrift), in Himmelweit (ed.), *The Collected Papers of Paul Ehrlich*, Vol. II, pp. 256–77 (English translation, pp. 278–97).

Ehrlich, Paul and Hans Sachs (1905), 'Über den Mechanismus der Antiamboceptorwirkung' (Berliner Klinische Wochenschrift), in Himmelweit (ed.), *The Collected Papers of Paul Ehrlich*, Vol. II, pp. 432–41.

Ehrlich, Paul and Hans Sachs (1909), 'Kritiker der Seitenkettentheorie im Lichte ihrer experimentellen und literarischen Forschung. Ein Kommentar zu den Arbeiten von Bang und Forssmann' (Münchener Medizinische Wochenschrift), in Himmelweit (ed.), *The Collected Papers of Paul Ehrlich*, Vol. II, pp. 448–63.

Ehrlich, Paul and Hans Sachs (1910), 'Ist die Ehrlichsche Seitenkettentheorie mit den tatsächlichen Verhältnissen vereinbar? Bemerkungen zu der II. Mitteilung von Bang und Forssmann' (Münchener Medizinische Wochenschrift), in Himmelweit (ed.), *The Collected Papers of Paul Ehrlich*, Vol. II, pp. 464–71.

Ehrlich, Paul and August Wassermann (1894), 'Über die Gewinnung der Diphtherie-Antitoxine aus Blutserum und Milch immunisirter Thiere' (Zeitschrift für Hygiene und Infektionskrankheiten), in Himmelweit (ed.), *The Collected Papers of Paul Ehrlich*, Vol. II, pp. 72–9.

Ehrlich, Paul, Hermann Kossel and August Wassermann (1894), 'Über Gewinnung und Verwendung des Diphtherieheilserums' (Deutsche Medizinische Wochenschrift), in Himmelweit (ed.), *The Collected Papers of Paul Ehrlich*, Vol. II, pp. 56–60.

Eldefrawi, M.E. and A.T. Eldefrawi (1972), 'Characterization and Partial Purification of the Acetylcholine Receptor from *Torpedo electroplax*', *Proceedings of the National Academy of Sciences*, 69: 1776–80.

Elkeles, Barbara (1985), 'Medizinische Menschenversuche gegen Ende des 19. Jahrhunderts und der Fall Neisser: Rechtfertigung und Kritik einer wissenschaftlichen Methode', *Medizinhistorisches Journal*, 20: 135–48.

Elliott, Thomas Renton (1904), 'On the Action of Adrenalin', *Journal of Physiology*, 31: xx–xxi.

Elliott, Thomas Renton (1905), 'The Action of Adrenalin', *Journal of Physiology*, 32: 401–67.

Elliott, Thomas Renton (1907), 'The Innervation of the Bladder and Urethra', *Journal of Physiology*, 35: 367–445.

Engelhardt, E. and O. Loewi (1930), 'Fermentative Azetylcholinspaltung im Blut und ihre Hemmung durch Physostigmin', *Archiv für experimentelle Pathologie und Pharmakologie*, 150: 1–13.

Erbslöh, Friedrich (1958), 'Muskelkrankheiten', in Gustav Bodechtel (ed.), *Differentialdiagnose neurologischer Krankheitsbilder* (Stuttgart: Georg Thieme), pp. 723–55.

Eulner, Hans-Heinz (1970), *Die Entwicklung der medizinischen Spezialfächer an den Universitäten des deutschen Sprachgebietes* (Studien zur Medizingeschichte des 19. Jahrhunderts, 4) (Stuttgart: Enke).

Ewald, Gottfried (1944), *Lehrbuch der Neurologie und Psychiatrie* (Munich, Berlin: Lehmann).

Eyer, Peter (2004), 'Walther-Straub-Institut für Pharmakologie und Toxikologie, Medizinische Fakultät der Ludwig-Maximilians-Universität München', in Athineos Philippu (ed.), *Geschichte und Wirken der pharmakologischen, klinisch-pharmakologischen und toxikologischen Institute im deutschssprachigen Raum* (Bruneck: Berenkamp), pp. 518–31.

Farber, Eduard (1981), 'Fischer, Emil Hermann', in *Dictionary of Scientific Biography*, Vol. 5, 10th edn (New York: Charles Scribner's Sons), pp. 1–5.

Feldberg, W. (1967), 'John Henry Gaddum 1900–1965', *Biographical Memoirs of Fellows of the Royal Society*, 13: 57–77.

Feldberg, W. (1970), 'Henry Hallett Dale', *Biographical Memoirs of Fellows of the Royal Society*, 16: 77–174.

Feldberg, W., A. Fessard and D. Nachmansohn (1940), 'The Cholinergic Nature of the Nervous Supply to the Electric Organ of the Torpedo (*Torpedo marmorata*)', *Journal of Physiology*, 97: 3P–4P.

Flad-Schnorrenberg, Beatrice (1980), 'Eine Cella europäischer Cultur. Die Stazione Zoologica in Neapel', in Hans-Reiner Simon (ed.), *Anton Dohrn und die Zoologische Station Neapel* (Frankfurt: Verlag Edition Erbrich), pp. 153–6 (reprint of an article in the *Frankfurter Allgemeine Zeitung*, No. 58, 8 March 1980).

Fletcher, W.M. (1926), 'John Newport Langley. In Memoriam', *Journal of Physiology*, 61: 1–27.

Fletcher, W.M. (1927), 'John Newport Langley – 1852–1925', *Proceedings of the Royal Society of London*, Series B, 101: xxxiii–xli.

Forst, A.W. (1974), 'Walther Straub 100 Jahre', *Münchener Medizinische Wochenschrift*, 116: 1171–4.

Foster, M. and J.N. Langley (1899), *A Course of Elementary Practical Physiology and Histology*, 7th edn, edited by J. N. Langley and L. E. Shore (London: Macmillan).

Franken, Franz Hermann (1994), *Friedrich Theodor Frerichs (1819–1885). Leben und hepatologisches Werk* (Freiburg: Falk Foundation).

French, R. (1975), *Antivivisection and Medical Science in Victorian Society* (Princeton, NJ: Princeton University Press).

Fröhlich, A. and O. Loewi (1908), 'Untersuchungen zur Physiologie und Pharmakologie des autonomen Nervensystems. I. Mitteilung über die Wirkung der Nitrite und des Atropins', *Archiv für experimentelle Pathologie und Pharmakologie*, 59: 34–56.

Fühner, H. (1907), 'Curarestudien. I. Die periphere Wirkung des Guanidins', *Archiv für experimentelle Pathologie und Pharmakologie*, 58: 1–49.

Gaddum, J.H. (1926), 'The Action of Adrenalin and Ergotamine on the Uterus of the Rabbit', *Journal of Physiology*, 61: 141–50.

Gaddum, J.H. (1937), 'The Quantitative Effects of Antagonistic Drugs', *Journal of Physiology*, 89: 7P–9P.

Gaddum, J.H. (1941), 'Prof. A.J. Clark, F.R.S.', *Nature*, 148: 189–90.

Gaddum, J.H. (1943), 'Introductory Address. Part I. Biological Aspects: the Antagonism of Drugs', *Transactions of the Faraday Society*, 39: 323–32.

Gaddum, J.H. (1964), 'Foreword', in E.J. Ariëns (ed.), *Molecular Pharmacology: the Mode of Action of Biologically Active Compounds*, 2 vols (New York: Academic Press), vol. 1, p. xi.

Geison, G.L. (1973), 'Langley, John Newport', in *Dictionary of Scientific Biography*, ed. C.C. Gillispie (New York: Charles Scribner's Sons), vol. 8, pp. 14–19.

Geison, G.L. (1978), *Michael Foster and the Cambridge School of Physiology. The Scientific Enterprise in Late Victorian Society* (Princeton, NJ: Princeton University Press).

Geison, G.L. (1981), 'Otto Loewi', in *Dictionary of Scientific Biography*, ed. C.C. Gillispie (New York: Charles Scribner's Sons), vol. 7, pp. 451–7.

Gieryn, Thomas F. (1995), 'Boundaries of Science', in Sheila Jasanoff, Gerald E. Markle, James C. Petersen and Trevor Pinch (eds), *Handbook of Science and Technology Studies* (Thousand Oaks, London and New Dehli: Sage), pp. 393–443.

Goldsmith, Margaret (1934), 'Paul Ehrlich', in Hector Bolitho (ed.), *Twelve Jews* (London: Rich & Cowan Ltd.), pp. 65–81.

Golinski, Jan (1998), *Making Natural Knowledge: Constructivism and the History of Science* (Cambridge: Cambridge University Press).

Goodman, Jordan (2000), 'Pharmaceutical Industry', in Roger Cooter and John Pickstone (eds), *Medicine in the 20th Century* (Amsterdam: Harwood), pp. 141–54.

Gradmann, Christoph (1998), 'Leben in der Medizin: Zur Aktualität von Biographie und Prosopographie in der Medizingeschichte', in Norbert Paul and Thomas Schlich (eds), *Medizingeschichte: Aufgaben, Probleme, Perspektiven* (Frankfurt and New York: Campus), pp. 243–65.

Gradmann, Christoph (2000), 'Money and Microbes: Robert Koch, Tuberculin and the Foundation of the Institute for Infectious Diseases in Berlin in 1891', *History and Philosophy of the Life Sciences*, 22: 59–79.

Gradmann, Christoph (2001), 'Robert Koch and the Pressures of Scientific Research: Tuberculosis and Tuberculin', *Medical History*, 45: 1–32.

Gradmann, Christoph (2004), 'A Harmony of Illusions: Clinical and Experimental Testing of Robert Koch's Tuberculin 1890–1900', *Studies in History and Philosophy of Biological and Biomedical Sciences*, 35: 465–81.

Gradmann, Christoph (2005), *Krankheit im Labor. Robert Koch und die medizinische Bakteriologie* (Göttingen: Wallstein).

Greene, Charles Wilson (1914), *Handbook of Pharmacology* (New York: William Wood and Company).

Greiling, Walter (1954), *Im Banne der Medizin. Paul Ehrlich. Leben und Werk* (Düsseldorf: Econ).

Gremels, Hans (1947), 'Walther Straub †', *Archiv für experimentelle Pathologie und Pharmakologie*, 204: 2–12.

Grmek, M.D. (1973), *Raissonement Expérimental et Recherches Toxicologiques chez Claude Bernard* (Geneva: Librairie Droz).

Gruber, Max (1901a), 'Zur Theorie der Antikörper. I. Ueber die Antitoxin-Immunität', *Münchener Medizinische Wochenschrift*, 48: 1827–30, 1880–4.

Gruber, Max (1901b), 'Zur Theorie der Antikörper. II. Ueber Bakteriolyse und Haemolyse', *Münchener Medizinische Wochenschrift*, 48: 1924–7, 1965–8.

Gruber, Max (1903a), 'Toxin und Antitoxin. Bemerkungen zu Ehrlichs "Entgegnung auf Grubers Replik"', *Münchener Medizinische Wochenschrift*, 50: 2297.

Gruber, Max (1903b), 'Neue Früchte der Ehrlichschen Toxinlehre', *Wiener Klinische Wochenschrift*, 16: 791–3.

Gruber, Max (1903c), 'Toxin und Antitoxin. Eine Replik auf Herrn Ehrlichs Entgegnung', *Münchener Medizinische Wochenschrift*, 50: 1825–8.

Gruber, Max and Cl. Freiherr v. Pirquet (Referent: M. Gruber) (1903), 'Toxin und Antitoxin', *Münchener Medizinische Wochenschrift*, 50, pp. 1193–6.

Guillebeau, Julie (1977), 'Dr. Ahlquist Receives Pharmacology's Highest Honor for His Clinical Research Work on Alpha-Beta Receptors', *MCG Today*, 6 (Spring) (2): 4–5.

Gunn, J.A. (1931), *An Introduction to Pharmacology and Therapeutics*, 2nd edn (London and Edinburgh: Oxford University Press).

Gunn, J.A. (1932), *An Introduction to Pharmacology and Therapeutics*, 3rd edn (London and Edinburgh: Oxford University Press).

Gunn, J.A. (1949), 'Dixon, Walter Ernest (1870–1931)', *Dictionary of National Biography 1931–1940* (Oxford and London: Oxford University Press, Geoffrey Cumberlege), p. 231.

Günther, Oswin (1954), 'Immunitätstheorien. Von der Seitenkettentheorie zur Fließbandtheorie', in *Zum 100. Geburtstage Paul Ehrlichs und zum Wiederaufbau des Paul-Ehrlich-Instituts*, Arbeiten aus dem Paul-Ehrlich-Institut, dem Georg-Speyer-Haus und dem Ferdinand-Blum-Institut zu Frankfurt a.M., Issue 51 (Stuttgart: Gustav Fischer), pp. 68–107.

Haffner, F. (1944), 'Walther Straub', *Münchener Medizinische Wochenschrift*, 91: 343–4.

Hagist, C. and L.J. Kotlikoff (2007), 'Health Care Spending: What the Future Will Look Like', NCPA Policy Report No. 286, June, www.ncpa.org/pub/st/st286.

Hagner, Michael (2001), 'Ansichten der Wissenschaftsgeschichte', in *Ansichten der Wissenschaftsgeschichte* (Frankfurt am Main: Fischer), pp. 7–39.

Handovsky, Hans, Eveline Du Bois-Reymond and Christa Marie von Strantz (1923), 'Beeinflussung der Vitalität von Protozoen durch chemische Reize, gemessen an der Teilungsgeschwindigkeit', *Archiv für Experimentelle Pathologie und Pharmakologie*, 100: 273–87.

Hannaway, Caroline and Ann La Berge (1998), *Constructing Paris Medicine* (Clio Medical, vol. 50) (Amsterdam, Atlanta/GA: Rodopi).

Hardy, Anne I. (2006), 'Paul Ehrlich und die Serumproduzenten: Zur Kontrolle des Diphtherieserums in Labor und Fabrik', *Medizinhistorisches Journal*, 41: 51–84.

Harig, G. (1974), *Bestimmung der Intensität im medizinischen System Galens. Ein Beitrag zur theoretischen Pharmakologie, Nosologie und Therapie in der Galenischen Medizin* (Berlin: Akademie-Verlag).

Harrington, Anne (1996), *Reenchanted Science: Holism and German Culture from Wilhelm II to Hitler* (Princeton, NJ: Princeton University Press).

Harwood, Jonathan (1993), *Styles of Scientific Thought: the German Genetics Community 1900–1933* (Chicago and London: University of Chicago Press).

Harwood, Jonathan (2000), 'The Rise of the Party-Political Professor? Changing Self-Understandings among German Academics, 1890–1933', in Doris Kaufmann (ed.), *Geschichte der Kaiser-Wilhelm-Gesellschaft im*

Nationalsozialismus. Bestandsaufnahme und Perspektiven der Forschung, vol. 1 (Göttingen: Wallstein), pp. 21–45.

Häusler, H. and O. Loewi (1925), 'Zur Frage der Wirkungsweise des Insulins. I. Mitteilung. Insulin und die Glucoseverteilung zwischen flüssigen und nicht-flüssigen Systemen', *Pflügers Archiv für die gesamte Physiologie des Menschen und der Tiere*, 210: 238–79.

Häusler, H. and O. Loewi (1926), 'Untersuchungen über Diabetes und Insulinwirkung. V. Mitteilung', *Pflügers Archiv für die gesamte Physiologie des Menschen und der Tiere*, 214: 370–9.

Hawgood, B.J. (2002), 'Professor Chen-Yuan Lee, MD (1915–2001), Pharmacologist: Snake Venom Research at the Institute of Pharmacology, National Taiwan University', *Toxicon*, 40: 1065–72.

Hazell Clark, D. (1985), *Alfred Joseph Clark 1885-1941: a Memoir* (Glastonbury: C. & J. Clark Ltd. Archives).

Healy, David (2002), *The Creation of Psychopharmacology* (Cambridge, MA and London: Harvard University Press).

Hecht, Selig (1931), 'Die physikalische Chemie und die Physiologie des Sehaktes', in L. Asher and K. Spiro (eds), *Ergebnisse der Physiologie*, Vol. 32 (Munich: J.F. Bergmann), pp. 243–390.

Heidelberger, Michael (1996), 'Fechner, Gustav Theodor', in *Deutsche Biographische Enzyklopädie*, Vol. 3 (Darmstadt: Wissenschaftliche Buchgesellschaft), p. 238.

Heubner, Wolfgang (1944), 'Zu Walther Straub's 70. Geburtstag', *Klinische Wochenschrift*, 23: 139.

Heymann, Bruno (1928), 'Zur Geschichte der Seitenkettentheorie Paul Ehrlichs', *Klinische Wochenschrift*, 7: 1257–60.

Hill, A.V. (1909), 'The Mode of Action of Nicotine and Curare, Determined by the Form of the Contraction Curve and the Method of Temperature Coefficients', *Journal of Physiology*, 39: 361–73.

Hill, A.V. (1910), 'The Possible Effects of the Aggregation of the Molecules of Haemoglobin on its Dissociation Curves', *Journal of Physiology*, 40: iv–vii.

Hill, A.V. (1926), 'Lecture I: the Physical Environment of the Living Cell', in H.H. Dale, J.C. Drummond, L.J. Henderson and A. V. Hill, *Lectures on Certain Aspects of Biochemistry* (London: University of London Press), pp. 253–80.

Hill, A.V. (1965), 'The Mechanism of Muscular Contraction', in *Nobel Lectures Including Presentation Speeches and Laureates' Biographies. Physiology or Medicine 1922–1941*, edited by the Nobel Foundation (Amsterdam: Elsevier), pp. 10–23.

Himmelweit, Fred (ed.) (1956), with the assistance of Martha Marquardt, under the editorial direction of Sir Henry Dale, *The Collected Papers of Paul Ehrlich in four Volumes Including a Complete Bibliography*, Vol. I, Histology, Biochemistry and Pathology (London, New York: Pergamon).

Himmelweit, Fred (ed.) (1957), with the assistance of Martha Marquardt, under the editorial direction of Sir Henry Dale, *The Collected Papers of Paul Ehrlich in four Volumes Including a Complete Bibliography*, Vol. II, Immunology and Cancer Research (London, New York, Paris: Pergamon).

Himmelweit, Fred (ed.) (1960), with the assistance of Martha Marquardt, under the editorial direction of Sir Henry Dale, *The Collected Papers of Paul Ehrlich in four Volumes Including a Complete Bibliography*, Vol. III, Chemotherapy (London, Oxford, New York, Paris: Pergamon).

Hoch, W., J. McConville, S. Helms, J. Newsom-Davis, A. Melms and A. Vincent (2001), 'Autoantibodies to the Receptor Tyrosine Kinase MuSK in Patients with Myasthenia Gravis without Acetylcholine Receptor Antibodies', *Nature Medicine*, 7: 365–8.

Hoff, P. (1995), 'Kraepelin. Clinical Section – Part I', in German E. Berrios and Roy Porter (eds), *A History of Clinical Psychiatry* (London: Athlone), pp. 261–79.

Houlston, T. (1784), *Observations on Poisons; and on the Use of Mercury in the Cure of Obstinate Dysenteries* (London: Printed for R. Baldwin, by H. Reynell).

Howell, Joel D. (1995), *Technology in the Hospital. Transforming Patient Care in the Early Twentieth Century* (Baltimore, MD: Johns Hopkins University Press).

Hubenstorf, Michael (2001), 'Gruber, Max (von)', in Wolfgang U. Eckart and Christoph Gradmann (eds), *Ärzte Lexikon. Von der Antike bis zur Gegenwart*, 2nd edn (Berlin, Heidelberg: Springer), pp. 140–1.

Hucho, F. and J.-P. Changeux (1973), 'Molecular Weight and Quaternary Structure of the Cholinergic Receptor Protein Extracted by Detergents from *Electrophorus Electricus* Electric Tissue', *FEBS Letters*, 38: 11–15.

Ing, H.R. (1943), 'Chemical Constitution and Pharmacological Action', *Transactions of the Faraday Society*, 39: 372–80.

Joannovics, Georg (1915), 'Paul Ehrlich 1854–1915', *Wiener Klinische Wochenschrift*, 28: 937–42.

Jokl, Ernst (1954), 'Paul Ehrlich – Man and Scientist', *Bulletin of the New York Academy of Medicine*, 30: 968–75.

Jütte, Robert (1996), 'Wo alles anfing: Deutschland', in Martin Dinges (ed.), *Weltgeschichte der Homöopathie. Länder-Schulen-Heilkundige* (Munich: Beck), pp. 19–47.

Karlin, A. (1974), 'The Acetylcholine Receptor: Progress Report', *Life Sciences*, 14: 1385–415.

Kass Wenger, Nanette and Lowell M. Greenbaum (1984), 'From Adrenoceptor Mechanisms to Clinical Therapeutics: Raymond Ahlquist, PhD, 1914–1983', *Journal of the American College of Cardiology*, 3: 419–21.

Katz, B. (1978), 'Archibald Vivian Hill', *Biographical Memoirs of Fellows of the Royal Society*, 24: 71–149.

Katz, B. and R. Miledi (1970), 'Membrane Noise Produced by Acetylcholine', *Nature*, 226: 962–3.

Katz, B. and R. Miledi (1971), 'Further Observations on Acetylcholine Noise', *Nat. New Biol.*, 232: 124–6.

Katz, B. and R. Miledi (1973), 'The Characteristics of "End-Plate Noise" Produced by Different Depolarizing Drugs', *Journal of Physiology*, 230: 707–17.

Katz, B. and S. Thesleff (1957), 'A Study of the "Desensitization" Produced by Acetylcholine at the Motor End-Plate', *Journal of Physiology*, 138: 63–80.

Kay, Lily E. (1993), *The Molecular Vision of Life: Caltech, the Rockefeller Foundation, and the Rise of the new Biology* (New York and Oxford: Oxford University Press).

Keyser, P.T. (1997), 'Science and Magic in Galen's Recipes (Sympathy and Efficacy)', in A. Debru (ed.), *Galen on Pharmacology: Philosophy, History and Medicine* (Leiden: Brill), pp. 175–98.

Klasen, Eva-Maria (1984), 'Die Diskussion über eine "Krise" der Medizin in Deutschland zwischen 1925 und 1935', Med. Thesis, Mainz.

Klein, E., J.N. Langley and E.A. Schäfer (1883), 'On the Cortical Areas Removed from the Brain of a Dog, and from the Brain of a Monkey', *Journal of Physiology*, 4: 231–47.

Klose, F. (1954), 'Paul Ehrlich und Emil v. Behring. Zur hundertjährigen Wiederkehr ihrer Geburtstage am 14. und 15. März 1954', *Deutsche Medizinische Wochenschrift*, 79: 425–7.

Knight, David (2003), 'Sympathy, Attraction and Elective Affinity', *BSÉA XVII–XVIII*, 56: 21–30.

Koch, Richard (1924), 'Vorwort', in Martha Marquardt (ed.), *Paul Ehrlich als Mensch und Arbeiter. Erinnerungen aus dreizehn Jahren seines Lebens (1902–1915)* (Stuttgart, Berlin, Leipzig: DVA), pp. 3–15.

Kohler, Robert E. (1994), *Lords of the Fly: Drosophila Genetics and the Experimental Life* (Chicago: University of Chicago Press).

Kohler, Robert E. (1999), 'Moral Economy, Material Culture, and Community in Drosophila Genetics', in Mario Biagioli (ed.), *The Science Studies Reader* (New York and London: Routledge), pp. 243–57.

Kretschmer, W. (1907), 'Dauernde Blutdrucksteigerung durch Adrenalin und über den Wirkungsmechanismus des Adrenalins', *Archiv für experimentelle Parthologie und Pharmakologie*, 57: 423–37.

Krügel, Rainer (1984), *Friedrich Martius und der konstitutionelle Gedanke* (Marburger Schriften zur Medizingeschichte, Vol. 11) (Frankfurt/M. and Bern: Lang).

Kubalek, E., S. Ralston, J. Lindstrom and N. Unwin (1987), 'Location of Subunits within the Acetylcholine Receptor by Electron Image Analysis of Tubular Crystals from *Torpedo marmorata*', *Journal of Cellular Biology*, 105: 9–18.

Kuschinski, G. (1968), 'The Influence of Dorpat on the Emergence of Pharmacology as a Distinct Discipline', *Journal of the History of Medicine and Allied Sciences*, 23: 258–71.

Kuyer, A. and I.A. Wijsenbeek (1913), 'Über Entgiftungserregung und Entgiftungshemmung', *Pflügers Archiv für die gesamte Physiologie des Menschen und der Tiere*, 154: 16–38.

La Torre, J.L., G.S. Lunt and E. De Robertis (1970), 'Isolation of a Cholinergic Proteolipid Receptor from Electric Tissue', *Proceedings of the National Academy of Sciences*, 65 (3): 716–20.

Langley, J.N. (1875a), 'Preliminary Notice of Experiments on the Physiological Action of Jaborandi', *British Medical Journal*, 1875/I: 241–2.

Langley, J.N. (1875b), 'On the Physiological Action of Jaborandi', *Proceedings of the Cambridge Philosophical Society*, April: 402–4.

Langley, J.N. (1875c), 'The Action of Jaborandi on the Heart', *Journal of Anatomy and Physiology*, 10: 187–201.

Langley, J.N. (1876), 'The Action of Pilocarpin on the Sub-Maxillary Gland of the Dog', *Journal of Anatomy and Physiology*, 11: 173–80.

Langley, J.N. (1878), 'On the Physiology of the Salivary Secretion. Part II. On the Mutual Antagonism of Atropin and Pilocarpin, having Especial Reference to their Relations in the Sub-Maxillary Gland of the Cat', *Journal of Physiology*, 1: 339–69.

Langley, J.N. (1880), 'On the Antagonism of Poisons', *Journal of Physiology*, 3: 11–21.

Langley, J.N. (1883a), 'The Structure of the Dog's Brain', *Journal of Physiology*, 4: 248–85.

Langley, J.N. (1883b), 'Report on the Parts Destroyed on the Right Side of the Brain of the Dog Operated on by Prof. Goltz', *Journal of Physiology*, 4: 286–309.

Langley, J.N. (1898), 'The Salivary Glands', in E.A. Schäfer (ed.), *Text-Book of Physiology*, 2 vols (Edinburgh: Pentland), vol. 1, pp. 475–530.

Langley, J.N. (1899), 'Presidential Address, Section I – Physiology', *Report of British Association for 1899*, pp. 881–92.

Langley, J.N. (1900), 'The Sympathetic and Other Related Systems of Nerves', in E.A. Schäfer (ed.), *Text-Book of Physiology*, 2 vols (Edinburgh: Pentland), vol. 2, pp. 616–96.

Langley, J.N. (1901a), 'On the Stimulation and Paralysis of Nerve-Cells and of Nerve-Endings. Part I', *Journal of* Physiology, 27: 224–36.

Langley, J.N. (1901b), 'Observations on the Physiological Action of Extracts of the Supra-Renal Bodies', *Journal of Physiology*, 27: 237–56.

Langley, J.N. (1903a), 'The Autonomic Nervous System', Presidential Address, *Brain*, 26, part I: 1–26.

Langley, J.N. (1903b), 'Das sympathische und verwandte nervöse Systeme der Wirbeltiere (autonomes nervöses System)', *Ergebnisse der Physiologie*, 2, part 2: 818–72.

Langley, J.N. (1905a), 'The Autonomic Nerves', *Nederlandsch Tijdschrift voor Geneeskunde*, 16: 1013–30.

Langley, J.N. (1905b), 'On the Reaction of Cells and of Nerve-Endings to Certain Poisons, Chiefly as Regards the Reaction of Striated Muscle to Nicotine and to Curari', *Journal of Physiology*, 33: 374–413.

Langley, J.N. (1906a), 'Croonian Lecture, 1906 – On Nerve Endings and on Special Excitable Substances in Cells', *Proceedings of the Royal Society of London*, series B, 78: 170–94.

Langley, J.N. (1906b), 'Über Nervenendigungen und spezielle rezeptive Substanzen in Zellen', *Zentralblatt für Physiologie*, 20: 290–1.

Langley, J.N. (1907a), 'Sir Michael Foster. In Memoriam', *Journal of Physiology*, 35: 233–46.

Langley, J.N. (1907b), 'Nouvelles Observations sur la Nature Non-Spécifique des Terminaisons Nerveuses Motrices et sur l'Existence de Radicules "Réceptives" dans le Muscle', *Archives Internationales de Physiologie*, 5: 115–18.

Langley, J.N. (1907–09), 'On the Contraction of Muscle, Chiefly in Relation to the Presence of "Receptive" Substances', *Journal of Physiology*, 36: 347–84; 37: 165–212, 285–300; 39: 235–95.

Langley, J.N. (1909/10), 'On the Contraction of Muscle, Chiefly in Relation to the Presence of "Receptive" Substances. Part IV. The Effect of Curari and of Some Other Substances on the Nicotine Response of the Sartorius and Gastrocnemius Muscles of the Frog', *Journal of Physiology*, 39: 235–95.

Langley, J.N. (1910), 'Note on the Action of Nicotine and Curare on the Receptive Substance of the Frog's Rectus Abdominis Muscle', *Journal of Physiology*, 40: lviii–lix.

Langley, J.N. (1914), 'The Antagonism of Curare and Nicotine in Skeletal Muscle', *Journal of Physiology*, 48: 73–108.

Langley, J.N. (1915), 'Walter Holbrook Gaskell, 1847–1914', *Proceedings of the Royal Society of London*, series B, 88: xxvii–xxxvi.

Langley, J.N. (1918), 'Persistence of the Central Somatic Effect of Strychnine after a Large Dose of Nicotine', *Journal of Physiology*, 52: xliv–xlv.

Langley, J.N. (1921), *The Autonomic Nervous System. Part I* (Cambridge: W. Heffer & Sons) [no second part published].

Langley, J.N. and W.L. Dickinson (1889), 'On the Local Paralysis of Peripheral Ganglia, and on the Connexion of Different Classes of Nerve Fibres with them', *Proceedings of the Royal Society of London*, 46: 423–31.

Langley, J.N. and W.L. Dickinson (1890a), 'Pituri and Nicotine', *Journal of Physiology*, 11: 265–306.

Langley, J.N. and W.L. Dickinson (1890b), 'Action of Various Poisons upon Nerve-Fibres and Peripheral Nerve-Cells', *Journal of Physiology*, 11: 501–27

Langley, J.N. and A.S. Grünbaum (1890), 'On the Degeneration Resulting from Removal of the Cerebral Cortex and Corpora Striata in the Dog', *Journal of Physiology*, 11: 606–28.

Langley, J.N. and R. Magnus (1905), 'Some Observations of the Movements of the Intestine before and after Degenerative Section of the Mesenteric Nerves', *Journal of Physiology*, 33: 34–51.

Langley, J.N. and C.S. Sherrington (1884), 'Secondary Degeneration of Nerve Tracts following Removal of the Cortex of the Cerebrum in the Dog', *Journal of Physiology*, 5: 49–65.

'Lasker Laurels for Medical Researchers' (1976), *Medical World News*, 29 November: 48–50.

Lawrence, Christopher and George Weisz (eds) (1998), *Greater than the Parts: Holism in Biomedicine, 1920–1950* (New York and Oxford: Oxford University Press).

Lawrence, Christopher and George Weisz (1998), 'Medical Holism: the Context', in *Greater than the Parts: Holism in Biomedicine, 1920–1950* (New York and Oxford: Oxford University Press), pp. 1–22.

Lazarus, Adolf (1922), *Paul Ehrlich* (Meister der Heilkunde, Vol. 2) (Wien, Berlin, Leipzig and München: Ricola).

Leake, C.D. (1975), *An Historical Account of Pharmacology to the 20th Century* (Springfield, Ill: C.C. Thomas).

Lee, C.Y. and C.C. Chang (1966), 'Modes of Actions of Purified Toxins from Elapid Venoms on Neuromuscular Transmission', *Mem. Inst. Butantan*, 33 (2): 555–72.

Lee, C.Y., L.F. Tseng and T.H. Chiu (1967), 'Influence of Denervation on Localization of Neurotoxins from Clapid Venoms in Rat Diaphragm', *Nature*, 215(106): 1177–8.

Lembeck, Fred and Wolfgang Giere (1968), *Otto Loewi. Ein Lebensbild in Dokumenten* (Berlin, Heidelberg and New York: Springer).

Lembeck, Fred and Walter Kukovetz (1985), 'Otto Loewi', in Kurt Freisitzer, Walter Höflechner, Hans-Ludwig Holzer and Wolfgang Mantl (eds), *Tradition und Herausforderung. 400 Jahre Universität Graz* (Graz: Akademische Druck- und Verlagsanstalt), pp. 265–76.

Lembeck, Fred, Peter Holzer and Bernhard A. Peskar (2004), 'Institut für Experimentelle und Klinische Pharmakologie, Medizinische Fakultät der Karl-Franzens-Universität Graz', in Athineos Philippu (ed.), *Geschichte und Wirken der pharmakologischen, klinisch-pharmakologischen und toxikologischen Institute im deutschsprachigen Raum* (Bruneck: Berenkamp), pp. 256–69.

Lennig, Petra (1994a), 'Die Entwicklung des Grundkonzeptes der Psychophysik durch Gustav Theodor Fechner – eine spezielle Lösungsvariante des philosophisch tradierten Leib-Seele-Problems?', *Naturwissenschaft. Technik. Medizin.*, 2: 159–74.

Lennig, Petra (1994b), *Von der Metaphysik zur Psychophysik. Gustav Theodor Fechner (1801–1887). Eine ergobiographische Studie* (Beiträge zur Geschichte der Psychologie, Vol. 8) (Frankfurt/M. and Berlin: Peter Lang).

Lenoir, Timothy (1988), 'A Magic Bullet: Research for Profit and the Growth of Knowledge in Germany around 1900', *Minerva*, 26: 66–88.

Lenoir, Timothy (1997), *Instituting Science: the Cultural Production of Scientific Disciplines* (Stanford, CA: Stanford University Press).

Lesch, John E. (1984), *Science and Medicine in France: the Emergence of Experimental Physiology, 1790–1855* (Cambridge, MA: Harvard University Press).

Lesch, John E. (2007), *The First Miracle Drugs: How the Sulfa Drugs Transformed Medicine* (Oxford and New York: Oxford University Press).

Levitan, E.D., P.R. Schofield, D.R. Burt, L.M. Rhee, W. Wisden, M. Kohler, N. Fujita, H. Rodriguez, F.A. Stephenson, M.G. Darlison, E.A. Barnard and P.H. Seeburg (1988), 'Structural and Functional Basis for $GABA_A$ Receptor Heterogeneity', *Nature*, 335: 76–9.

Levy, Bernhard and Raymond P. Ahlquist (1957), 'Inhibition of the Adrenergic Depressor Response', *Journal of Pharmacology and Experimental Therapeutics*, 121: 414–20.

Levy, Bernhard and Raymond P. Ahlquist (1960), 'Blockade of the *Beta* Adrenergic Receptors', *Journal of Pharmacology and Experimental Therapeutics*, 130: 334–9.

Levy, Bernhard and Raymond P. Ahlquist (1961), 'An Analysis of Adrenergic Blocking Activity', *Journal of Pharmacology and Experimental Therapeutics*, 133: 202–10.

Lewandowsky, M. (1899), 'Ueber die Wirkung des Nebennierenextractes auf die glatten Muskeln, im Besonderen des Auges', *Pflüger's Archiv für Anatomie und Physiologie*, 1899: 360–6.

Liebenau, Jonathan (1990), 'Paul Ehrlich as a Commercial Scientist and Research Administrator', *Medical History*, 34: 65–78.

Linton, Derek S. (2005), *Emil von Behring: Infectious Disease, Immunology, Serum Therapy* (Philadelphia: American Philosophical Society).

Little, R.C. (1987), 'Raymond P. Ahlquist (1914–1983)', *Clinical Cardiology*, 10: 583–4.

Loewi, Otto (1902a), 'Ueber Eiweisssynthese im Thierkörper', *Archiv für experimentelle Pathologie und Pharmakologie*, 48: 303–30.

Loewi, Otto (1902b), 'Untersuchungen zur Physiologie und Pharmakologie der Nierenfunktion', *Archiv für experimentelle Pathologie und Pharmakologie*, 48: 410–38.

Loewi, Otto (1908), 'Über vasoconstrictorische Fasern in der chorda tympani', *Archiv für experimentelle Pathologie und Pharmakologie*, 59: 64–70.

Loewi, Otto (1910), 'Pharmakologie und Klinik', *Wiener Klinische Wochenschrift*, 23: 273–8.

Loewi, Otto (1921), 'Über humorale Übertragbarkeit der Herznervenwirkung. I. Mitteilung', *Pflügers Archiv für die gesamte Physiologie des Menschen und der Tiere*, 189: 239–42.

Loewi, Otto (1922), 'Über humorale Übertragbarkeit der Herznervenwirkung. II. Mitteilung', *Pflügers Archiv für die gesamte Physiologie des Menschen und der Tiere*, 193: 201–13.

Loewi, Otto (1924a), 'Über humorale Übertragbarkeit der Herznervenwirkung. III. Mitteilung', *Pflügers Archiv für die gesamte Physiologie des Menschen und der Tiere*, 203: 408–12.

Loewi, Otto (1924b), 'Über humorale Übertragbarkeit der Herznervenwirkung. IV. Mitteilung', *Pflügers Archiv für die gesamte Physiologie des Menschen und der Tiere*, 204: 361–7.

Loewi, Otto (1924c), 'Über humorale Übertragbarkeit der Herznervenwirkung. V. Mitteilung. Die Übertragbarkeit der negativ chrono- und dromotropen Vaguswirkung', *Pflügers Archiv für die gesamte Physiologie des Menschen und der Tiere*, 204: 629–40.

Loewi, Otto (1926), 'Ueber die Wirkung des Insulins und des Insulin-Antagonisten des diabetischen Blutes', *Wiener Klinische Wochenschrift*, 39: 1074–6.

Loewi, Otto (1927a), 'Über Strukturfixierung der Glucose und ihre Bedeutung für das Glucoseschicksal', *Die Naturwissenschaften*, 15: 93–4.

Loewi, Otto (1927b), 'Glykämin und Insulin', *Klinische Wochenschrift*, 6: 2169–76.

Loewi, Otto (1929), 'Bemerkungen zur Rezeptorenfunktion im Verdauungskanal', *Deutsche Medizinische Wochenschrift*, 55: 1751–2.

Loewi, Otto (1930), 'Bemerkungen zur Rezeptorenfunktion im Verdauungskanal', *Klinische Wochenschrift*, 9: 1089.

Loewi, Otto (1931), 'Über "neuro-humorale" Auslösung im Organismus', *Jahreskurse für ärztliche Fortbildung*, 22: 1–11.

Loewi, Otto (1937a), 'Die chemische Übertragung der Nervenwirkung. Nobelvortag', 12 December 1936, Stockholm, in *Les Prix Nobel en 1936* (Stockholm: Norstedt & Söner), pp. 1–14.

Loewi, Otto (1937b), 'Die chemische Uebertragung der Nervenwirkung', *Schweizerische Medizinische Wochenschrift*, 18: 850–5.

Loewi, Otto (1960), 'An Autobiographic Sketch', *Perspectives in Biology and Medicine*, Autumn: 3–25.

Loewi, Otto (1965), 'The Chemical Transmission of Nerve Action', in *Nobel Lectures Including Presentation Speeches and Laureates' Biographies. Physiology or Medicine 1922–1941* (Amsterdam, London and New York: Elsevier, for the Nobel Foundation), pp. 416–29.

Loewi, Otto and E. Navratil (1924), 'Über humorale Übertragbarkeit der Herznervenwirkung. VI. Mitteilung. Der Angriffspunkt des Atropins', *Pflügers Archiv für die gesamte Physiologie des Menschen und der Tiere*, 206: 123–34.

Loewi, Otto and E. Navratil (1926a), 'Über humorale Übertragbarkeit der Herznervenwirkung. X. Mitteilung. Über das Schicksal des Vagusstoffs', *Pflügers Archiv für die gesamte Physiologie des Menschen und der Tiere*, 214: 678–88.

Loewi, Otto and E. Navratil (1926b), 'Über humorale Übertragbarkeit der Herznervenwirkung. XI. Mitteilung. Über den Mechanismus der Vaguswirkung von Physostigmin und Ergotamin', *Pflügers Archiv für die gesamte Physiologie des Menschen und der Tiere*, 214: 689–96.

Loewi, Otto and Grete Singer (1924), 'Über die Wirkung des Jods auf die Atmung isolierter Zellen', *Wiener Medizinische Wochenschrift*, 74: 328–32.

Loewi, Otto and J. Solti (1922), 'Über die Wirkung von Pilocarpin und Atropin auf den isolierten Krötenmuskel und ihre Abhängigkeit von der Ionen-Mischung', *Klinische Wochenschrift*, 1: 2046.

Löwy, Ilana (ed.) (1993a), *Medicine and Change: Historical and Sociological Studies of Medical Innovation* (Montrouge, Paris and London: Colloques INSERM, John Libbey Eurotext Ltd.).

Löwy, Ilana (1993b), 'Introduction: Medicine and Change', in *Medicine and Change: Historical and Sociological Studies of Medical Innovation* (Montrouge, Paris and London: Colloques INSERM, John Libbey Eurotext Ltd.), pp. 1–20.

Lucae, Christian (1998), *Homöopathie an deutschsprachigen Universitäten. Die Bestrebungen zu ihrer Institutionalisierung von 1812 bis 1945* (Quellen und Studien zur Homöopathiegeschichte, Vol. 4) (Heidelberg: Karl F. Haug Verlag).

Luchsinger, B. (1877), 'Die Wirkungen von Pilocarpin und Atropin auf die Schweissdrüsen der Katze. Ein Beitrag zur Lehre vom doppelseitigen Antagonismus zweier Gifte', *Pflüger's Archiv für Anatomie und Physiologie*, 15: 482–92.

Lupton, Deborah (1994), *Medicine as Culture. Illness, Disease and the Body in Western Societies* (London, Thousand Oaks and New Delhi: Sage).

Maehle, A.-H. (1987), *Johann Jakob Wepfer (1620–1695) als Toxikologe* (Aarau, Frankfurt/Main, Salzburg: Verlag Sauerländer).

Maehle, A.-H. (1999), *Drugs on Trial: Experimental Pharmacology and Therapeutic Innovation in the Eighteenth Century* (Amsterdam and Atlanta, GA: Rodopi).

Maehle, A.-H. (2002), 'Sistemi e metodi terapeutici', in *Storia della scienza*, editor-in-chief Sandro Petruccioli, vol. 10: L'Età dei Lumi (Rome: Istituto della Enciclopedia Italiana), pp. 726–36.

Maehle, A.-H. (2004a), 'Historische Grundlagen des Rezeptor-Konzepts in der Pharmakologie', *Gesnerus*, 61: 57–76.

Maehle, A.-H. (2004b), ' "Receptive Substances": John Newport Langley (1852–1925) and his Path to a Receptor Theory of Drug Action', *Medical History*, 48: 153–74.

Maehle, A.-H. (2005), 'The Quantification and Differentiation of the Drug Receptor Theory, *c.* 1910–1960', *Annals of Science*, 62: 479–500.

Maehle, A.-H., C.-R. Prüll and R.F. Halliwell (2002), 'The Emergence of the Drug Receptor Theory', *Nature Reviews. Drug Discovery*, 1: 637–41.

Magnus, O. (2002), *Rudolf Magnus: Physiologist and Pharmacologist 1873–1927* (Dordrecht: Kluwer Academic Publishers).

Magnus, R. (1908), 'Kann man den Angriffspunkt eines Giftes durch antagonistische Giftversuche bestimmen?' *Pflüger's Archiv für die gesamte Physiologie*, 123: 99–112.

Maienschein, Jane (1990), 'Cell Theory and Development', in Richard C. Olby et al. (eds), *Companion to the History of Modern Science* (London, New York: Routledge), pp. 357–73.

Malkin, Harold M. (1993), *Out of the Mist: the Foundation of Modern Pathology and Medicine during the Nineteenth Century* (Berkeley: Vesalius Books).

Malkin, Harold M. (2003), 'Concept of Acid-Base Balance in Medicine', *Annals of Clinical & Laboratory Science*, 33: 337–44.

Marquardt, Martha (1924), *Paul Ehrlich als Mensch und Arbeiter. Erinnerungen aus dreizehn Jahren seines Lebens (1902–1915)* (Stuttgart, Berlin and Leipzig: DVA).

Marquardt, Martha (1949), *Paul Ehrlich* (London: Heinemann).

Marshall, Liz (1995), 'Paul Ehrlich, 1854–1915. German Bacteriologist and Immunologist', in Emily J. McMurray et al. (eds), *Notable Twentieth-Century Scientists*, vol. 1 (New York and London: Gale Research Inc.), pp. 564–7.

Martius, Friedrich (1923), 'Das Arndt-Schulzsche Grundgesetz', *Münchener Medizinische Wochenschrift*, 70: 1005–6.

Matthews, Bryan H.C. (1931), 'The Response of a Single End Organ', *Journal of Physiology*, 71: 64–110.

Maulitz, Russell C. (1978), 'Rudolf Virchow, Julius Cohnheim, and the Program of Pathology', *Bulletin of the History of Medicine*, 52: 162–82.

Mazumdar, Pauline M.H. (1974), 'The Antigen-Antibody Reaction and the Physics and Chemistry of Life', *Bulletin of the History of Medicine*, 48: 1–21.

Mazumdar, Pauline M.H. (1989), 'The Template Theory of Antibody Formation and the Chemical Synthesis of the Twenties', in Pauline M.H. Mazumdar (ed.), *Immunology 1930–1980: Essays on the History of Immunology* (Toronto: Wall & Thompson), pp. 13–32.

Mazumdar, Pauline M.H. (1995), *Species and Specificity: an Interpretation of the History of Immunology* (Cambridge: Cambridge University Press).

McCormmach, Russell (1982), *Night Thoughts of a Classical Physicist* (Cambridge, MA and London: Harvard University Press).

'MCG Professor to Receive Hunter Memorial Citation' (1973), *The Augusta Chronicle*, 27 December.

Meunier, J.-C., R.W. Olsen, A. Menez, P. Fromageot, P. Boquet and J.-P. Changeux (1972), 'Some Physical Properties of the Cholinergic Receptor Protein from *Electrophorus electricus* Revealed by a Tritiated α-Toxin from *Naja nigricollis* Venom', *Biochemistry*, 11: 1200–10.

Meyer, Hans (1899), 'Zur Theorie der Alkoholnarkose. Erste Mittheilung. Welche Eigenschaften der Anästhetica bedingt ihre narkotische Wirkung?' *Archiv für experimentelle Pathologie und Pharmakologie*, 42: 109–18.

Meyer, Hans (1901), 'Zur Theorie der Alkoholnarkose. Dritte Mittheilung: Der Einfluss wechselnder Temperatur auf Wirkungsstärke und Theilungscoefficient der Narcotica', *Archiv für experimentelle Pathologie und Pharmakologie*, 44: 338–46.

Meyer, Hans Horst (1922), 'Schmiedebergs Werk', *Archiv für experimentelle Pathologie und Pharmakologie*, 92: 1–17.

Meyer, Oskar B. (1908), 'Versuche mit Kokain-Adrenalin und Andolin an überlebenden Blutgefäßen', *Zeitschrift für Biologie*, 50: 93–112.

Michaelis, Leonor (1919), 'Zur Erinnerung an Paul Ehrlich: Seine wiedergefundene Doktor-Dissertation', *Die Naturwissenschaften*, 7: 165–8.

Miledi, R. (1966), 'Miniature Synaptic Potentials in Squid Nerve Cells', *Nature*, 212: 1240–2.

Miledi, R. (1973), 'Transmitter Release Induced by Injection of Calcium Ions into Nerve Terminals', *Proceedings of the Royal Society of London*, B 183: 421–5.

Miledi, R., P. Molinoff and L.T. Potter (1971), 'Isolation of the Cholinergic Receptor Protein of Torpedo Electric Tissue', *Nature*, 229 (5286): 554–7.

Mishina, M., T. Takahashi, T. Takai, M. Kurasaki, K. Fukuda and S. Numa (1985), 'Role of Acetylcholine Receptor Subunits in Gating of the Channel, *Nature*, 318: 538–43.

Moran, Neil C. and Marjorie E. Perkins (1958), 'Adrenergic Blockade of the Mammalian Heart by a Dichloro Analogue of Isoproterenol', *Journal of Pharmacology and Experimental Therapeutics*, 124: 223–37.

Moran, Neil C. and Marjorie E. Perkins (1961), 'An Evaluation of Adrenergic Blockade of the Mammalian Heart', *Journal of Pharmacology and Experimental Therapeutics*, 133: 192–201.

Moraw, Peter (1988), 'Vom Lebensweg des deutschen Professors', *Forschung. Mitteilungen der DFG*, no. 4: 3.

Morrell, Jack B. (1972), 'The Chemist Breeders: the Research Schools of Liebig and Thomson', *Ambix* 19: 1–46.

Morrell, Jack B. (1993), 'W.H. Perkin, Jr., at Manchester and Oxford: From Irwell to Isis', *Osiris*, 2nd series, vol. 8, Research Schools: Historical Reappraisals, pp. 104–26.

Mould, Richard F. (1993), *A Century of X-Rays and Radioactivity in Medicine: With Emphasis on Photographic Records of the Early Years* (Bristol and Philadelphia: Institute of Physics Publishing).

Moulin, Anne Marie (1991), *Le dernier langage de la médecine. Histoire de l'immunologie de Pasteur au Sida* (Paris: Presses Universitaires de France).

Müller, Irmgard (1976), 'Die Geschichte der Zoologischen Station in Neapel von der Gründung durch Anton Dohrn (1872) bis zum Ersten Weltkrieg und ihre Bedeutung für die Entwicklung der modernen biologischen Wissenschaften', Habilitation Thesis, Düsseldorf.

Müller-Jahncke, Wolf-Dieter and Christoph Friedrich (1996), *Geschichte der Arzneimitteltherapie* (Stuttgart: Deutscher Apotheker Verlag).

Münch, Ragnhild and Stefan S. Biel (1998), 'Expedition, Experiment und Expertise im Spiegel des Nachlasses von Robert Koch', *Sudhoffs Archiv*, 82: 1–29.

Nana Djiepmo, Judith (2005), 'Über den Einfluss der Nachlastbedingungen auf Eigenschaften des stunned Myokard. Eine Studie an blutperfundierten, isolierten Kaninchenherzen', Med. Thesis, Düsseldorf.

Nanda, T.C. (1931), 'Factors Influencing the Response of Plain Muscle to Drugs', *Quarterly Journal of Experimental Physiology*, 21: 141–6.

Neukirch, P. (1913), 'Physiologische Wertbestimmung am Dünndarm (nebst Beiträgen zur Wirkungsweise des Pilokarpins)', *Pflügers Archiv für die gesamte Physiologie des Menschen und der Tiere*, 147: 153–70.

Nickerson, M. (1956), 'Receptor Occupancy and Tissue Response', *Nature*, 178: 697–8.

Noda, M., H. Takahashi, T. Tanabe, M. Toyosato, Y. Furutani, T. Hirose, M. Asai, S. Inayama, T. Miyata and S. Numa (1982), 'Primary Structure of Alpha-Subunit Precursor of *Torpedo californica* Acetylcholine Receptor Deduced from cDNA Sequence', *Nature*, 299: 793–7.

Noda, M., H. Takahashi, T. Tanabe, M. Toyosato, S. Kikyotani, T. Hirose, M. Asai, H. Takashima, S. Inayama, T. Miyata and S. Numa (1983a), 'Primary Structures of Beta- and Delta-Subunit Precursors of *Torpedo californica* Acetylcholine Receptor Deduced from cDNA Sequences', *Nature*, 301: 251–5.

Noda, M., H. Takahashi, T. Tanabe, M. Toyosato, S. Kikyotani, Y. Furutani, T. Hirose, H. Takashima, S. Inayama, T. Miyata and S. Numa (1983b), 'Structural Homology of *Torpedo californica* Acetylcholine Receptor Subunits', *Nature*, 302: 528–32.

Nurmand, Leo (2004), 'Pharmakologisches Laboratorium und Pharmakologisches Institut, Medizinische Fakultät der Universität Dorpat (Tartu)', in Athineos Philippu (ed.), *Geschichte und Wirken der pharmakologischen, klinisch-phar makologischen und toxikologischen Institute im deutschssprachigen Raum* (Bruneck: Berenkamp), pp. 151–9.

Obituary (1915a), 'Professor Paul Ehrlich', *British Medical Journal*, 2: 349.

Obituary (1915b), 'Wirklicher Geheimrat Paul Ehrlich', *Lancet*, 2: 525–6.

Oliver, G. and E.A. Schäfer (1895), 'The Physiological Effects of Extracts of the Suprarenal Capsules', *Journal of Physiology*, 18: 230–76.

Olmsted, J.M.D. (1939), *Claude Bernard, Physiologist* (London: Cassell).

Olsen, R.W., J.-C. Meunier and J.-P. Changeux (1972), 'Progress in the Purification of the Cholinergic Receptor Protein from *Electrophorus electricus* by Affinity Chromatography', *FEBS Letters*, 28: 96–100.

Otis, Laura (2000), 'Arthur Conan Doyle: an Imperial Immune System, in *Membranes. Metaphors of Invasion in Nineteenth-Century Literature, Science, and Politics* (Baltimore and London: Johns Hopkins University Press), pp. 90–118.

Overton, E. (1899), 'Ueber die allgemeinen osmotischen Eigenschaften der Zelle, ihre vermutlichen Ursachen und ihre Bedeutung für die Physiologie', *Vierteljahrbuch der Naturforschergesellschaft Zürich*, 44: 88–135.

Overton, E. (1901), *Studien über die Narkose, zugleich ein Beitrag zur allgemeinen Pharmakologie* (Jena: Gustav Fischer).

Parascandola, J. (1974), 'The Controversy over Structure-Activity Relationships in the Early Twentieth Century', *Pharmacy in History*, 16(2): 54–63.

Parascandola, J. (1975), 'Arthur Cushny, Optical Isomerism, and the Mechanism of Drug Action', *Journal of the History of Biology*, 8: 145–65.

Parascandola, J. (1981), 'The Theoretical Basis of Paul Ehrlich's Chemotherapy', *Journal of the History of Medicine and Allied Sciences*, 36: 19–43.

Parascandola, J. (1982), 'A.J. Clark: Quantitative Pharmacology and the Receptor Theory', *Trends in Pharmacological Sciences*, 3: 421–3.

Parascandola, J. (1986), 'The Development of Receptor Theory', in M. J. Parnham and J. Bruinvels (eds), *Discoveries in Pharmacology*, 3 vols (Amsterdam: Elsevier), vol. 3, pp. 129–56.

Parascandola, J. (1992), *The Development of American Pharmacology: John J. Abel and the Shaping of a Discipline* (Baltimore: Johns Hopkins University Press).

Parascandola, J. and R. Jasensky (1974), 'Origins of the Receptor Theory of Drug Action', *Bulletin of the History of Medicine*, 48: 199–220.

Parnham, M.J. and J. Bruinvels (eds) (1983–86), *Discoveries in Pharmacology*, 3 vols (Amsterdam: Elsevier).

Partsch, K.J. (1980), *Die Zoologische Station in Neapel* (Göttingen: Vandenhoeck & Ruprecht).

Paton, W.D.M. (1961), 'A Theory of Drug Action Based on the Rate of Drug-Receptor Combination', *Proceedings of the Royal Society of London*, series B, 154: 21–69.

Patrick, J. and J. Lindstrom (1973), 'Autoimmune Responses to Acetylcholine Receptor', *Science*, 180: 871–2.

Perks, Robert (1990), *Oral History: an Annotated Bibliography* (London: British Library).

Perks, Robert and Alistair Thomson (1997), *The Oral History Reader* (London: Routledge).

Pickstone, John (ed.) (1992), *Medical Innovations in Historical Perspective* (London: Macmillan).

Plimmer, H.G. (1897), 'A Critical Summary of Ehrlich's Recent Work on Toxins and Antitoxins', *Journal of Pathology and Bacteriology*, 5: 489–98.

Powell, C.E. and Irvine H. Slater (1958), 'Blocking of Inhibitory Adrenergic Receptors by a Dichloro Analog of Isoproterenol', *Journal of Pharmacology and Experimental Therapeutics*, 122: 480–8.

Prichard, Brian (1964), 'Hypotensive Action of Pronethalol', *British Medical Journal*, 1: 1227–8.

Prichard, Brian (1978), 'Beta Adrenergic Blocking Drugs and the Treatment of High Blood Pressure', in Karl Heinrich Rahn and Alfred Schrey (eds), *Betablocker. 1. Betadrenol-Symposion Frankfurt 1977* (Munich, Vienna and Baltimore: Urban & Schwarzenberg), pp. 126–46.

Prichard, Brian (1979), 'β-Adrenoceptor-Blocking Agents in the Management of Hypertension', *Cardiology*, 64, Suppl. 1: 44–87.

Prichard, Brian and Peter M.S. Gillam (1964), 'The Use of Propranolol in the Treatment of Hypertension', *British Medical Journal*, 2: 725–7.

'Professor Carl Gerhardt †' (1902), *Lokal-Anzeiger*, 20(337), Berlin, 22 July: 1.

'Profile (R.P. Ahlquist)' (1975), *Intercom*, 5, 9 May.

Prüll, C.-R. (1998), 'Holism and German Pathology (1914–1933)', in Christopher Lawrence and George Weisz (eds), *Greater than the Parts: Holism in Biomedicine, 1920–1950* (New York and Oxford: Oxford University Press), pp. 46–67.

Prüll, C.-R. (2003a), *Medizin am Toten oder am Lebenden? Pathologie in Berlin und in London 1900 bis 1945* (Veröffentlichungen der Gesellschaft für Universitäts- und Wissenschaftsgeschichte, 5) (Basel: Schwabe-Verlag).

Prüll, C.-R. (2003b), 'Part of a Scientific Master Plan? Paul Ehrlich and the Origins of his Receptor Concept', *Medical History*, 47: 332–56.

Prüll, C.-R., A.-H. Maehle and R.F. Halliwell (2003), 'Drugs and Cells – Pioneering the Concept of Receptors', *Pharmacy in History*, 45: 18–30.

Pütter, August (1918), 'Studien zur Theorie der Reizvorgänge', *Pflügers Archiv für die gesamte Physiologie des Menschen und der Tiere*, 171: 201–61.

Quirke, Viviane (2005), 'From Evidence to Market: Alfred Spinks's 1953 Survey of new Fields for Pharmaceutical Research, and the Origins of ICI's Cardiovascular Programme', in Virginia Berridge and Kelly Loughlin (eds), *Medicine, the Market and the Mass Media: Producing Health in the Twentieth Century* (London, New York: Routledge), pp. 146–71.

Quirke, Viviane (2006), 'Putting Theory into Practice: James Black, Receptor Theory and the Development of the Beta-Blockers at ICI, 1958–1978', *Medical History*, 50: 69–92.

Quirke, Viviane (2007), *Collaboration in the Pharmaceutical Industry: Changing Relationships in Britain and France, ca. 1935–1965* (London and New York: Routledge).

Raftery, M.A., M.W. Hunkapiller, C.D. Strader and L.E. Hood (1980), 'Acetylcholine Receptor: Complex of Homologous Subunits', *Science*, 208: 1454–7.

Ramberg, Peter J. (2003), *Chemical Structure, Spatial Arrangement: the Early History of Stereochemistry, 1874–1914* (Aldershot: Ashgate).

Rang, H.P. and Lord Perry of Walton (1996), 'Sir William Drummond Macdonald Paton, C.B.E.', *Biographical Memoirs of Fellows of the Royal Society*, 42: 291–314.

Rapport, Richard (2005), *Nerve Endings: the Discovery of the Synapse* (New York and London: W.W. Norton & Company).

'Recognition Justly Due' (1977), *Augusta Chronicle*, editorial, 23 March.

Remane, Horst (1984), *Emil Fischer* (Leipzig: Teubner).

Rentz, Eduard (1929), 'Vom Mechanismus mehrphasiger Wirkungen. II. Teil: Zusammenfassung der Literatur und Versuch einer allgemeinen Wertung (Theorie) der Phasenwirkungen', *Archiv für experimentelle Pathologie und Pharmakologie,* 141: 183–227.

Reynolds, J.A. and A. Karlin (1978), 'Molecular Weight in Detergent Solution of Acetylcholine Receptor from *Torpedo californica'*, *Biochemistry,* 17: 2035–8.

Riddle, J.M. (1992), *Quid pro quo: Studies in the History of Drugs* (Aldershot: Ashgate).

Reynolds, G.P. (2007), 'The Impact of Pharmacogenetics on the Development and Use of Antipsychotic Drugs', *Drug Discovery Today,* 12: 953–9.

Rheinberger, Hans-Jörg and Michael Hagner (eds) (1993), *Die Experimentalisierung des Lebens: Experimentalsysteme in den biologischen Wissenschaften 1850/1950* (Berlin: Akademie-Verlag).

Robinson, J.D. (2001), *Mechanisms of Synaptic Transmission: Bridging the Gaps (1890–1990)* (Oxford: Oxford University Press).

Rolleston, Sir Humphrey (1931), 'Obituary Walter Ernest Dixon', *The Times,* 17 August.

Rossbach, M.J. (1879), 'Neue Studien über den physiologischen Antagonismus der Gifte', *Pflüger's Archiv für Anatomie und Physiologie,* 21: 1–38.

Rost, Eugen (1947), 'Walther Straub in Memoriam!', *Medizinische Klinik,* 42: 202–4.

Rubin, Lewis P. (1980), 'Styles in Scientific Explanation: Paul Ehrlich and Svante Arrhenius on Immunochemistry', *Journal of the History of Medicine and Allied Sciences,* 35: 397–425.

Rupke, N.A. (ed.) (1990), *Vivisection in Historical Perspective* (London: Routledge).

Sauerteig, L. (1996), 'Salvarsan und der "ärztliche Polizeistaat". Syphilistherapie im Streit zwischen Ärzten, pharmazeutischer Industrie, Gesundheitsverwaltung und Naturheilverbänden (1900–1927)', in M. Dinges (ed.), *Medizinkritische Bewegungen im Deutschen Reich (ca. 1870 – ca. 1933)* (Stuttgart: Franz Steiner Verlag), pp. 161–200.

Schäfer, E.A. (ed.) (1898/1900), *Text-Book of Physiology,* 2 vols (Edinburgh: Pentland).

Schild, H.O. and F.L. Rose (1976), 'Harry Raymond Ing', *Biographical Memoirs of Fellows of the Royal Society,* 22: 239–55.

Schleifer, L.A. and M.E. Eldefrawi (1974), 'Identification of the Nicotinic and Muscarinic Acetylcholine Receptors in Subcellular Fractions of Mouse Brain', *Neuropharmacology,* 13(1): 53–63.

Schlich, Thomas (1998), 'Wissenschaft: Die Herstellung wissenschaftlicher Fakten als Thema der Geschichtsforschung', in Norbert Paul and Thomas Schlich (eds), *Medizingeschichte: Aufgaben, Probleme, Perspektiven* (Frankfurt and New York: Campus), pp. 107–29.

Schlich, Thomas (2000), 'Linking Cause and Disease in the Laboratory: Robert Koch's Method of Superimposing Visual and "Functional" Representations of Bacteria', *History and Philosophy of the Life Sciences,* 22: 43–58.

Schmiedebach, Heinz-Peter (1999), 'The Prussian State and Microbiological Research – Friedrich Loeffler and his Approach to the "Invisible" Virus', *Archives of Virology,* 15 (Suppl.): 9–23.

Schmiedeberg, Oswald (1867), 'Ueber die quantitative Bestimmung des Chloroforms im Blute und sein Verhalten gegen dasselbe', *Archiv der Heilkunde*, 8: 273–320.

Schmiedeberg, Oswald (1912), 'Rudolf Buchheim, sein Leben und seine Bedeutung für die Begründung der wissenschaftlichen Arzneimittellehre und Pharmakologie', *Archiv für experimentelle Pathologie und Pharmakologie*, 67: 1–54.

Schofield, P.R., M.G. Darlison, N. Fujita, D.R. Burt, F.A. Stephenson, H. Rodriguez, L.M. Rhee, J. Ramachandran, V. Reale, T.A. Glencorse, P.H. Seeburg and E.A. Barnard (1987), 'Sequence and Functional Expression of the GABA$_A$ Receptor Shows a Ligand-Gated Receptor Superfamily', *Nature*, 328: 221–7.

Schulz, Hugo (1888), 'Ueber Hefegifte', *Pflügers Archiv für die gesamte Physiologie des Menschen und der Tiere*, 42: 517–41.

Schüttler, J. (ed.) (2003), *50 Jahre Gesellschaft für Anästhesiologie und Intensivmedizin. Tradition und Innovation* (Berlin, Heidelberg and New York: Springer).

Shanks, Robin G. (1984), 'The Discovery of Beta Adrenoceptor Blocking Drugs', in M.J. Parnham and J. Bruinvels (eds), *Discoveries in Pharmacology*, vol. 2: Haemodynamics, Hormones & Inflammation (Amsterdam: Elsevier), pp. 37–72.

Sherrington, C. (1953), 'Marginalia', in E. Ashworth Underwood (ed.), *Science, Medicine and History: Essays in Honour of Charles Singer* (London: Oxford University Press), vol. 2, pp. 545–53.

Silverstein, Arthur M. (1989), *A History of Immunology* (San Diego and New York: Academic Press).

Silverstein, Arthur M. (1999), 'Paul Ehrlich's Passion: the Origins of his Receptor Immunology', *Cellular Immunology*, 194: 213–21.

Silverstein, Arthur M. (2000), 'Pasteur, Pastorians, and the Dawn of Immunology: the Importance of Specificity', *History and Philosophy of the Life Sciences*, 22: 29–41.

Silverstein, Arthur M. (2002), *Paul Ehrlich's Receptor Immunology: the Magnificent Obsession* (San Diego and London: Academic Press).

Simon, H.-R. (ed.) (1980), *Anton Dohrn und die Zoologische Station Neapel* (Frankfurt: Verlag Edition Erbrich).

Singer, S.J. and J.L. Nicolson (1972), 'The Fluid Mosaic Model of the Structure of Cell Membranes', *Science*, 175 (23): 720–31.

Singer, Wendy (1997), *Creating Histories: Oral Narratives and the Politics of History-Making* (Oxford: Oxford University Press).

Slater, Irwin H. and C.E. Powell (1959), 'Some Aspects of Blockade of Inhibitory Adrenergic Receptors or Adrenoceptive Sites', *Pharmacological Revue*, 11: 462–3.

Sollmann, Torald (1901), *A Text-Book of Pharmacology and some Allied Sciences (Therapeutics, Materia Medica, Pharmacy, Prescription-writing, Toxicology etc.)* (Philadelphia, London: W.B. Saunders & Company).

Sollmann, Torald (1922), *A Manual of Pharmacology and its Application to Therapeutics and Toxicology* (Philadelphia and London: W.B. Saunders Co.).

Sollmann, Torald and Paul J. Hanzlik (1928), *An Introduction to Experimental Pharmacology* (Philadelphia, London: W.B. Saunders Co).

Stannard, J. (1961), 'Hippocratic Pharmacology', *Bulletin of the History of Medicine*, 35: 497–518.

Stanton, Jennifer (ed.) (2002), *Innovations in Health and Medicine. Diffusion and Resistance in the Twentieth Century* (London and New York: Routledge).

Stapleton, Melanie Patricia (1997), 'Sir James Black and Propranolol: the Role of the Basic Sciences in the History of Cardiovascular Pharmacology', *Texas Heart Institute Journal*, 24: 336–42.

Star, Susan L. and James R. Griesemer (1989), 'Institutional Ecology, "Translations" and Boundary Objects: Amateurs and Professionals in Berkeley's Museum of Vertebrate Zoology, 1907–39', *Social Studies of Science*, 19: 387–420.

Starke, Klaus (1998), 'A History of Naunyn-Schmiedeberg's Archives of Pharmacology', *Naunyn-Schmiedeberg's Archives of Pharmacology*, 358: 1–109.

Starke, Klaus (2004a), *Die Geschichte des Pharmakologischen Instituts der Universität Freiburg* (Berlin and Heidelberg: Springer).

Starke, Klaus (2004b), 'Institut für Experimentelle und Klinische Pharmakologie und Toxikologie, Medizinische Fakultät der Albert-Ludwigs-Universität Freiburg', in Athineos Philippu (ed.), *Geschichte und Wirken der pharmakologischen, klinisch-pharmakologischen und toxikologischen Institute im deutschssprachigen Raum* (Bruneck: Berenkamp), pp. 214–23.

Stein, M. (1997), 'La Thériaque chez Galien: sa Préparation et son Usage Thérapeutique', in A. Debru (ed.), *Galen on Pharmacology: Philosophy, History and Medicine* (Leiden: Brill), pp. 199–208.

Stephenson, R.P. (1956), 'A Modification of Receptor Theory', *British Journal of Pharmacology*, 11: 379–93.

Stille, G. (1994), *Der Weg der Arznei von der Materia Medica zur Pharmakologie. Der Weg von Arzneimittelforschung und Arzneitherapie* (Karlsruhe: G. Braun Fachverlage).

Storm van Leeuwen, W. (1924), 'A Possible Explanation for Certain Cases of Hypersensitiveness to Drugs in Men', *Journal of Pharmacology and Experimental Therapeutics*, 24: 25–32.

Straub, Walther (1897), 'Ueber die Bedingungen des Auftretens der Glykosurie nach der Kohlenmonoxydvergiftung', *Aepp.*, 38: 139–57.

Straub, Walther (1903), 'Quantitative Untersuchung des Eindringens von Alkaloiden in lebende Zellen', *Pflügers Archiv für die gesamte Physiologie des Menschen und der Tiere*, 98: 233–40.

Straub, Walther (1905), 'Mechanismus der Muskarinwirkung am Herzen und des Antagonismus Atropin-Muskarin', *Zentralblatt für Physiologie*, 19: 302–4.

Straub, Walther (1907), 'Zur chemischen Kinetik der Muskarinwirkung und des Antagonismus Muskarin-Atropin', *Pflügers Archiv für die gesamte Physiologie des Menschen und der Tiere*, 119: 127–51.

Straub, Walther (1908), *Gift und Organismus. Öffentliche Antrittsrede, gehalten am 26. Februar 1908 in der Universitäts-Aula Freiburg i. Br.* (Leipzig: Speyer & Kaerner).

Straub, Walther (1910), 'Quantitative Untersuchungen über den Chemismus der Strophanthinwirkung', *Biochemische Zeitschrift*, 28: 392–407.

Straub, Walther (1911), 'Über chronische Vergiftungen, speziell die chronische Bleivergiftung', *Deutsche Medizinische Wochenschrift*, 32: 1469–71.

Straub, Walther (1912), *Die Bedeutung der Zellmembran für die Wirkung chemischer Stoffe auf den Organismus*, Gesellschaft Deutscher Naturforscher und Ärzte, Verhandlungen 1912, Sonderabdruck (Leipzig: August Pries).

Straub, Walther (1915), 'Tetanustherapie mit Magnesiumsulfat. Erfahrungen am tetanuskranken Menschen bei intravenöser Einführung des Magnesiumsulfats', *Münchener Medizinische Wochenschrift*, 10: 341–2.

Straub, Walther (1916), 'Chemischer Bau und pharmakologische Wirksamkeit in der Digitalisgruppe', *Biochemische Zeitschrift*, 75: 132–44.

Straub, Walther (1919), 'Toxikologische Untersuchung des M. Fickerschen Gasbrandtoxins und Antitoxins', *Münchener Medizinische Wochenschrift* (1919): 89–91.

Straub, Walther (1920), 'Naturwissenschaften und Pharmakologie im medizinischen Unterricht', *Deutsche Medizinische Wochenschrift*, 9/11: 246–7, 270–1, 300–1.

Straub, Walther (1931a), *Lane Lectures on Pharmacology. V. General Pharmacology of Heavy Metals* (Stanford: Stanford University Press), pp. 60–71.

Straub, Walther (1931b), *Lane Lectures on Pharmacology. IV. Digitalis: Biochemistry* (Stanford: Stanford University Press), pp. 45–59.

Straub, Walther (1936), 'Versuche über den Vagusstoff', *Archiv für experimentelle Pathologie und Pharmakologie*, 182: 331–9.

Straub, Walther (1937), 'Mode of Action of Stimulating Alkaloids', *Proceedings of the Royal Society of London*, series B, 121: 584–7.

Straub, Walther (1938), 'Discussion on the Scope and Future of Teaching and Research in Pharmacology [comment on the opening statement of Henry Hallett Dale]', in *Kongressbericht III des XVI. Internationalen Physiologenkongresses, 14.–19. August 1938 in Zürich*. Nachträge (Basel: Birkhäuser & Cie).

Straub, Walther and August Amman (1940), 'Über die Wirkung des Nicotins und die Diätetik des Tabakgenusses', *Klinische Wochenschrift*, 19: 169–71.

Straub, Walther and Ludwig Krehl (1919), 'Über Verodigen (Gitalin)', *Deutsche Medizinische Wochenschrift*, 10: 281–3.

'Straub, Walther', in *Deutsche Biographische Enzyklopädie*, vol. 9 (Munich: Saur) 1998, pp. 571–2.

Stroomann, Gerhard (1960), *Aus meinem roten Notizbuch. Ein Leben als Arzt auf Bühlerhöhe* (Frankfurt/Main: Societäts-Verlag).

Symposium given for Dr. Ahlquist (1980), *Beeper* (The Medical College of Georgia) 1(43), 12 November.

Szöllösi-Janze, Margit (1998), *Fritz Haber 1868–1934. Eine Biographie* (Munich: Beck).

Szöllösi-Janze, Margit (2000), 'Lebens-Geschichte – Wissenschaftsgeschichte. Vom Nutzen der Biographie für Geschichtswissenschaft und Wissenschaftsgeschichte', *Berichte zur Wissenschaftsgeschichte*, 23: 17–35.

Tansey, E.M. (1989), 'The Wellcome Physiological Research Laboratories 1894–1904: The Home Office, Pharmaceutical Firms, and Animal Experiments', *Medical History*, 33: 1–41.

Tansey, E.M. (1990), 'The Early Scientific Career of Sir Henry Dale FRS (1875–1968)', PhD Thesis, London.

Tansey, E.M. (1991), 'Chemical Neurotransmission in the Autonomic Nervous System: Sir Henry Dale and Acetylcholine', *Clinical Autonomic Research*, 1: 63–72.

Tansey, E.M. (1995a), 'What's in a Name? Henry Dale and Adrenaline, 1906', *Medical History*, 39: 459–76.

Tansey, E.M. (1995b), 'Sir Henry Dale and Autopharmacology: the Role of Acetylcholine in Neurotransmission', in C. Debru (ed.), *Essays in the History of the Physiological Sciences* (Amsterdam: Rodopi), pp. 179–93.

Taylor, Judy (1976), 'Nobel Odds Called Good by Ahlquist', *Augusta Chronicle*.

Thiselton-Dyer, W.T. (1907), 'Michael Foster – a Recollection', *Cambridge Review*, 30 May: 439–40.

Thompson, Paul (1991), 'Oral History and the History of Medicine: a Review', *Social History of Medicine*, 4: 371–83.

Thompson, Paul (2000), *The Voice of the Past: Oral History* (Oxford: Oxford University Press).

Thompson, Paul and Robert Perks (1993), *An Introduction to the Use of Oral History in the History of Medicine* (Bristol: Doveton/The British National Sound Archive).

Timmermann, Carsten (2001), 'Constitutional Medicine, Neoromanticism, and the Politics of Antimechanism in Interwar Germany', *Bulletin of the History of Medicine*, 75: 717–39.

Timmermann, Carsten (2006), 'Hexamethonium, Hypertension and Pharmaceutical Innovation: the Transformation of an Experimental Drug in Post-war Britain', in Carsten Timmermann and Julie Anderson (eds), *Devices and Designs: Medical Technologies in Historical Perspective* (Basingstoke: Palgrave Macmillan), pp. 156–74.

Toyoshima, C. and N. Unwin (1988), 'Ion Channel of Acetylcholine Receptor Reconstructed from Images of Postsynaptic Membranes', *Nature*, 336: 247–50.

Traube, Isidor (1910), 'Die Theorie des Haftdrucks (Oberflächendrucks) und ihre Bedeutung für die Physiologie', *Pflügers Archiv für die gesamte Physiologie des Menschen und der Tiere*, 132: 511–38.

Traube, Isidor (1911), 'Die Theorie des Haftdrucks (Oberflächendrucks) und ihre Bedeutung für die Physiologie', *Pflügers Archiv für die gesamte Physiologie des Menschen und der Tiere*, 140: 109–34.

Traube, Isidor (1919), 'Die physikalische Theorie der Arzneimittel- und Giftwirkung', *Biochemische Zeitschrift*, 98: 177–96.

Travis, Anthony S. (1989), 'Science as the Receptor of Technology: Paul Ehrlich and the Synthetic Dyestuffs Industry', *Science in Context*, 3: 383–408.

Ueber den Bericht des Koch'schen Instituts für Infectionskrankheiten (1892) (Leipzig: Thieme).

Unwin, N., C. Toyoshima and E. Kubalek (1988), 'Arrangement of the Acetylcholine Receptor Subunits in the Resting and Desensitized States, Determined by Cryoelectron Microscopy of Crystallized *Torpedo* Postsynaptic Membranes', *Journal of Cellular Biology*, 107: 1123–38.

Valenstein, Elliot S. (2005), *The War of the Soups and the Sparks: the Discovery of Neurotransmitters and the Dispute over how Nerves Communicate* (New York: Columbia University Press).

Verney, E.B. and J. Barcroft (1941), 'Alfred Joseph Clark 1885–1941', *Obituary Notices of Fellows of the Royal Society*, 3: 969–84.

Vincent, A., J. Palace and D. Hilton-Jones (2001), 'Myasthenia gravis. Seminar', *Lancet*, 357: 2122–8.

Wald, George (1991), 'Selig Hecht', in *National Academy of Sciences. Biographical Memoirs*, vol. 6: 80–101.

Ward, Alan (1995), *Copyright Ethics and Oral History* (Colchester: Oral History Society, University of Essex).

Warner, John Harley (1997), *The Therapeutic Perspective: Medical Practice, Knowledge, and Identity in America, 1820–1885* (Princeton, NJ: Princeton University Press).

Wassermann, August von (1909), 'Paul Ehrlich', *Münchener Medizinische Wochenschrift*, 56: 245–7.

Wassermann, August von (1914), 'Die Seitenkettentheorie', in Apolant et al., *Paul Ehrlich* (Jena: Gustav Fischer), pp. 134–50.

Watson, G. (1966), *Theriac and Mithridatium: a Study in Therapeutics* (London: Wellcome Historical Library).

Weatherall, M. (1990), *In Search of a Cure: a History of Pharmaceutical Discovery* (Oxford: Oxford University Press).

Weindling, Paul (1992), 'Scientific Elites and Laboratory Organisation in Fin de Siècle Paris and Berlin: the Pasteur Institute and Robert Koch's Institue for Infectious Diseases Compared', in Andrew Cunningham and Perry Williams (eds), *The Laboratory Revolution in Medicine* (Cambridge: Cambridge University Press), pp. 170–88.

Wertheimer, Ernst and Hans Paffrath (1925), 'Beziehungen zwischen Permeabilität und Wirkung bei den Vertretern der Cholingruppe', *Pflügers Archiv für die gesamte Physiologie des Menschen und der Tiere*, 207: 254–68.

Whitla, William (1903), *Elements of Pharmacy, Materia Medica and Therapeutics*, 8th edn (London: Henry Renshaw).

Whitla, William (1910), *Elements of Pharmacy, Materia Medica and Therapeutics*, 9th edn (London: Ballière, Tindall and Cox).

Whitla, William (1915), *Elements of Pharmacy, Materia Medica and Therapeutics*, 10th edn (London: Ballière, Tindall and Cox).

Whitla, William (1923), *Elements of Pharmacy, Materia Medica and Therapeutics*, 11th edn (London: Ballière, Tindall and Cox).

Whittaker, V. P. (1998), 'Arcachon and Cholinergic Transmission', *Journal of Physiology* (Paris) 92: 53–7.

Whittaker, V.P., I.A. Michaelson and R.J.A. Kirkland (1964), 'The Separation of Synaptic Vesicles from Nerve-Ending Paricles ("Synaptosomes")', *Biochemical Journal*, 90: 293–303.

Wilson, A. and H.O. Schild (1961), *Applied Pharmacology (Clark)*, 9th edn (London: J. & A. Churchill).

Winau, Rolf (1987), *Medizin in Berlin* (Berlin and New York: Walter de Gruyter).

Witebsky, Ernst (1954), 'Ehrlich's Side-Chain Theory in the Light of Present Immunology', *Annals of the New York Academy of Sciences*, 59: 168–81.

Witham, A.C. (1977), 'Ahlquist, Basic Research, and the Practice of Medicine', Editorial, *Journal of the Medical Association of Georgia*, 66: 637–8.

Wittern, R. (1991), 'The Origins of Homoeopathy in Germany', in W.F. Bynum and V. Nutton (eds), *Essays in the History of Therapeutics* (Amsterdam: Rodopi), pp. 51–63.

Wolfe, Elin W., Clifford A. Barger and Saul Benison (2000), *Walter B. Cannon, Science and Society* (Cambridge, MA: Boston Medical Library in the Francis A. Countway Library of Medicine. Distributed by the Harvard University Press).

Index